JOHN BILLINGSLEY

John Billingsley of Ashwick Grove, portrait by Joseph Hutchinson
(Victoria Art Gallery, Bath & North East Somerset Council)

John Billingsley

1747–1811

Entrepreneur turned Agriculturalist

'equalled by few, excelled by none'

LIN THORLEY

THE HOBNOB PRESS

First published in the United Kingdom in 2024

by The Hobnob Press,
8 Lock Warehouse, Severn Road, Gloucester GL1 2GA
www.hobnobpress.co.uk

© Lin Thorley 2024

The Author hereby asserts her moral rights to be identified as the Author of the Work.

All rights reserved. No part of this publication may be reproduced, stored in a retrieval system, or transmitted in any form or by any means, electronic, mechanical, photocopying, recording or otherwise, without the prior permission of the publisher and copyright holder.

British Library Cataloguing in Publication Data
A catalogue record for this book is available from the British Library

ISBN 978-1-914407-79-6

Typeset in Adobe Garamond Pro, 11/14 pt
Typesetting and origination by John Chandler

Front cover: Portrait of John Billingsley, by Joseph Hutchinson, reproduced by courtesy of the Victoria Art Gallery, Bath and North East Somerset.
Back cover: Title page of the General View of the Agriculture of the County of Somerset, by John Billingsley, second edition, 1795; Ashwick Grove, from the Auction Catalogue of 1937, Oakhill and Ashwick Local History Group collection.

Contents

Foreword		*vii*
Preface		*ix*
Introduction		*xi*
1	Ashwick – Background and Context	1
2	John Billingsley's Life – A Brief Synopsis, with a Table of Key Events	9 13
3	The Wool Trade	15
4	Turnpikes	23
5	The Oakhill Brewery	33
6	Agriculture and the Bath & West – Part 1	48
7	Agriculture and the Bath & West – Part 2	59
8	Enclosures	74
9	Land Acquisition	88
10	Navigable Canals	96
11	Mining	107
12	Water Management and the Somerset Levels	115
13	Ashwick Grove	130
14	General View of Agriculture in Somerset	142
15	An Essay on Waste Lands	158
16	Lord Waldegrave's Steward	172
17	Last Years and Legacy	182
Abbreviations		197
References		197
Selected Bibliography		207
Acknowledgements		208
Index		209

*To my husband David, my children
and grandchildren,
with my love*

Foreword

A BIOGRAPHY OF JOHN BILLINGSLEY is long overdue. A nonconformist farmer and entrepreneur from the Mendips, he became one of the most influential of the agricultural 'improvers' of the late eighteenth century, whilst no one individual can have had a greater impact on the landscape of Somerset.

Lin Thorley's thoroughly-researched biography does this remarkable man justice in all his many guises: wool trader, brewery owner, progressive farmer, turnpike trustee, pillar of the early Bath and West Society and author of the 1794 'General View of Agriculture in the County of Somerset' which tells us so much about farming in the county at a time of almost revolutionary change.

'He drained Sedgemoor; he inclosed Mendip; he wrote the Agricultural Survey of Somerset', declared Sir Benjamin Hobhouse, in delivering Billingsley's eulogy at the Bath and West Society's Annual Meeting in 1812, and he was scarcely exaggerating. To Billingsley, and his fellow enthusiast for drainage, Richard Locke of Burnham, we owe the green gridiron of the Somerset Levels; whilst it was Billingsley's leadership on enclosure and farm design which has given us the Mendip landscape of neat farmsteads and dry-stone walls.

In promulgating his ideas, Billingsley was fortunate to be able to work hand in glove with the Bath and West Society, founded in 1777 by Edmund Rack, a Norfolk quaker who shared Billingsley's innovative, science-based approach, to manufacturing as well as to agriculture, and who would become a great friend. Lin Thorley is quite clear that Billingsley was not a founder member of the Society, but he soon became very actively involved, winning the 'premiums' that it offered for, in his case, growing carrots, potatoes, cabbages and, most famously, his double-furrow plough. He was one of the most prolific contributors to the Society's Journals, which have proved to be a rich source of information on his innovations and ideas.

But if agricultural improvement was, as Lin Thorley puts it, 'the main focus of Billingsley's working life', it was by no means the only one. I hadn't realised, for example, what a central figure he had been in the Shepton Mallet 'spinning jenny' riots of 1776. He was one of a group of clothiers responsible for installing the machine which the rioters smashed, and it was Billingsley who tried to reason with them before the troops were brought in. He must have been brave, as well as persuasive, although in this case the local cloth workers were not to be persuaded, leading to the clothing industry moving decisively, and very damagingly for Shepton Mallet, northwards.

His involvement with the famous Oakhill Brewery seems to have been more commercially driven than born of any great interest in improving the technology of brewing. Again, he wasn't one of the founders, but his influence helped the business grow from a small rural concern in a Somerset backwater into a mega-brewery which, a century or so later, would be producing more stout than the Guinness company!

John Billingsley's character comes through from these pages, as well as his achievements. He obviously enjoyed a good dispute – about the respective merits of oxen and horses for ploughing, for example – and he could be a touch arrogant when it came to dismissing the habits of lesser mortals – his description of farming on the Somerset Levels, pre-drainage, being a classic example. 'The possession of a cow or two, with a hog and a few geese, naturally exalts the peasant in his own conception, sauntering after his cattle, he acquires a habit of indolence. Day-labour becomes disgusting; the aversion increases with indulgence; and at length the sale of a half-fed cow, or hog, furnishes the means of adding intemperance to idleness.' Harsh, maybe, but probably with more than a grain of truth. I also like the comment which Lin Thorley quotes about the relative merits of dairy farming and arable farming. Billingsley reckoned that dairying must be more profitable 'because dairy farmers are much more likely to pay their rents on time'!

The late eighteenth and early nineteenth centuries were a period of dramatic change in the Somerset countryside. This book chronicles those changes, and pays deserved tribute to a man who was one of their prime movers. It is a fascinating read, which sheds much new light on Billingsley and his times.

Anthony Gibson, OBE, MA (Oxon), FRAgS
Chair, Archive and Library Committee, Royal Bath and West Society

Preface

Standing at my window I can almost see the site of John Billingsley's home, Ashwick Grove. It lies hidden, down among the trees in a steep-sided valley at the far end of the fields, wrapped in Mendip mist. These days, sadly, the house is in ruins. But the place has a very special atmosphere, one of the things which inspired me to begin investigating the man said to have built it.

Mentioning the name 'John Billingsley' in a historical context may well get little response, even among historians. Many will not have heard of him. Others may know that he wrote a book on agriculture – or that he owned a brewery. Very few will know more. Once I began to look into the original records I realised that Billingsley is much underestimated. Part of the reason for this is, I believe, that beyond a few bare facts remarkably little has been written about him, in either scholarly or popular publications – and since his own time there are so many important parts of his life story that have never been mentioned at all (and many misconceptions reported by casual observers). Probably the best-known article is in Robin Atthill's *Old Mendip* from 1967, but it is only a short chapter.[1] On the more academic side Michael Williams' *The Draining of the Somerset Levels* and some of his papers from the 1970s analyse a limited area of Billingsley's work, but are little known outside the world of historical geography.[2,3] There has been no appraisal of Billingsley's overall work, either academic or popular. Yet he was important in his day, especially in the south-west of England. Atthill, Williams and others among the few who have mentioned Billingsley more recently have generally seen him as both able and influential in a variety of fields, particularly in agriculture.[4]

British agriculture underwent fundamental change in the eighteenth century, especially in its methods. Billingsley can be seen as an exemplar of the 'new farmer', of the sort generally known as an 'improver'. He was more capable and more shrewd than most; plus his innovations were on a broader scale. This meant he was at the forefront of the type that helped bring about agricultural change nationally. He was therefore chosen in the 1790s by the newly established Board of Agriculture to write about current agricultural practice in Somerset, to contribute to a series which helped to spread the word about current methods, suggested improvements and aided change. Today, the little that is known about eighteenth-century agricultural attitudes and practice in Somerset comes largely from that book: *General View of Agriculture in the County of Somerset with Observations on the Means of its Improvement* (1794).[5][*] The ideas he wrote about, there and elsewhere need exploration.

While a case can be made for attention to Billingsley for his capabilities in agriculture alone he was also involved in a remarkable variety of other activities, detailed within. Though he may not be considered the most pre-eminent of his time in any particular area, his breadth of accomplishment is unusual. It seemed to me that research on his almost forgotten life and work was long overdue.

Lin Thorley
17 July 2024

---

[*] The full title of Billingsley's book is abridged to '*General View..*' from here on.

Introduction

John Billingsley spent most of his life in the small parish of Ashwick, in Somerset's Mendip Hills. He was born about 1747 (the exact date is uncertain) into a well-known nonconformist family. At the time of his birth they were of relatively modest means but well educated and more intellectual than most. By the prime of his life Billingsley himself was thought of as a well-to-do gentleman. By the time of his death he was widely known in progressive circles as a practical agriculturalist, Encloser and drainage expert who paid great attention to cost effectiveness.* Relatively early in his career he moved his main financial focus from commerce into investing the proceeds in a brewery and in land, then turned his attention to cultivating and improving over 4,000 acres, most of it on the Mendip Hills. However, while he was progressive in many areas, his views on society and the poor, for example, were by contrast conservative.

A glance at the contents page of this book will show the multiple projects Billingsley became involved in other than agriculture: he was an extremely energetic man with a voracious appetite for work. Studying his various activities illuminates so many different aspects of the eighteenth century environment. On a personal level he was seen by his contemporaries as an amiable and dependable character, a man of integrity, highly thought of by his wide circle of friends and colleagues.

Despite his other interests, from early adulthood Billingsley's most serious work became increasingly focussed on agriculture. He published a number of articles, papers and reports on various subjects, all related to it in some way. His best-known work, the book on agriculture mentioned above, was first published in 1794, a second more important edition appearing in 1795. Many of the methods he advocated are still in use today, while some were, as he said, 'ridiculed' at the time. Through his active membership of the Bath and West Society he was in touch with many of the main agricultural

---

* A capital 'E' is used throughout for Enclosures referring to acts of parliament, as opposed to other enclosures, or the process of enclosing.

experts of the day. 'He Enclosed Mendip! He Drained the Levels!' was the claim just after his death.

John Billingsley died at Ashwick Grove in 1811, at the age of 64. Ashwick was a small world and he made few journeys beyond Bath and its environs, seldom venturing as far as London and making only two known trips further afield. But his ideas certainly travelled. My attention to Billingsley took me on a journey. I began to learn about his circle of friends and associates and about the many other areas he was involved in apart from agriculture. This book is intended to be a broader than usual biography: to include Billingsley's life, his work and times, and the circles he moved in. It is aimed at those with an interest in agricultural history or historical biography, whether they be professional historians or simply interested readers. Those readers concerned with one of the many other areas he was active in may also find the book useful for filling in some of the many gaps in the record or for the viewpoint it offers.

Due to Billingsley often being engaged in several different projects simultaneously, presenting the material chronologically could easily become confusing to the reader. Therefore, the decision was made early in the writing process that each topic would be treated separately (as much in sequence as reasonably possible), with a brief general description of his life and a timeline of his key projects included within chapter 2. This means the majority of chapters are relatively self-contained.

As most readers (especially those who are not professional historians) are likely to have much greater knowledge and interest in some areas than others, a very basic explanation is given as background to each – easily skipped. In addition, there are brief notes on the more important of the many other people who appear in the text. Readers who are chiefly interested in agriculture may wish to concentrate mainly on chapters 1 (Background and Context); 6 and 7 (Agriculture and the Bath and West Society); 8 (Enclosures); 11 (Water Management and the Somerset Levels); 14 (*General View...*); and 15 ('An Essay on Waste Lands').

An immediate and on-going difficulty in preparing this book has been the gaping holes in the available historical resources. The record is silent on significant areas of Billingsley's life. He left no papers: no correspondence, no diary, only one published book and a number of articles, almost all on agriculture. And as he had a finger in so many different pies the documents relating to his various activities are scattered and sporadic. For these reasons the text is inevitably uneven in its treatment of his numerous projects and of his life as a whole, which hopefully will not detract from its usefulness.

The information given here is founded on thorough original research, using primary sources wherever possible, a large amount of it previously unpublished. Sources are referenced where appropriate, so that those with a professional interest in the subject may easily find corroboration or further information. In spite of my efforts to be thorough, as this is the first published biography of Billingsley it is inevitable there will be errors of omission and commission – mistakes all my own – but which I hope will spur others to correct and bring further attention to this remarkable man.

Note
References are given numbers within the text and listed at the end of the book. Short additional notes appear at the foot of the page where appropriate.

I
Ashwick – Background and Context

To understand Billingsley's life and achievements a good place to start is the context – in terms of the time, place and circumstances – into which he was born.

Billingsley was a Mendip man. Ashwick is a small rural parish in the eastern part of the Mendip Hills, about six miles from the city of Wells and fifteen miles from Bath, the nearest town being Shepton Mallet just under three miles away. But in many senses Ashwick then was a world away from these centres of population, in the rural uplands. In Billingsley's time, as now, Ashwick itself was little more than a hamlet surrounded by farmland;

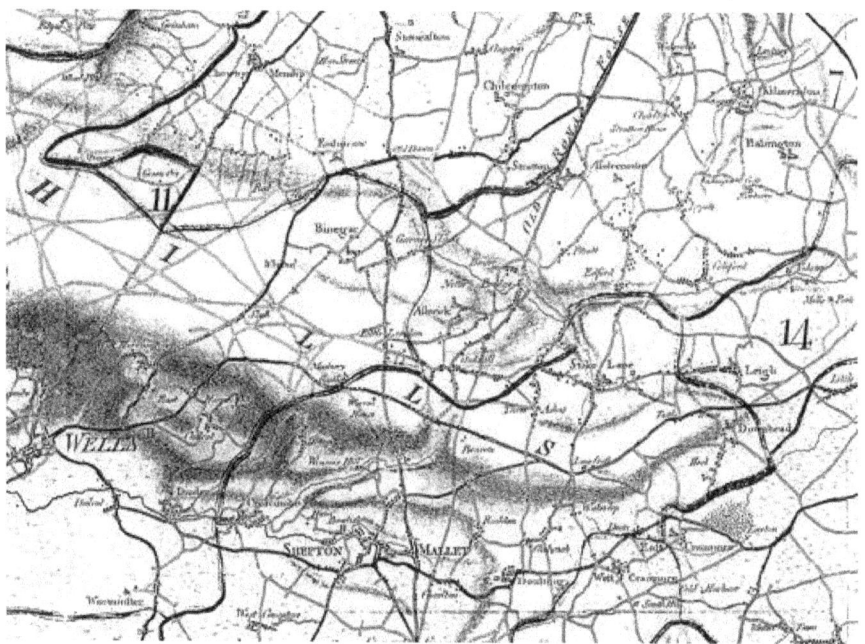

Detail from Day and Masters map of the County of Somerset 1782[1]
Ashwick is in the centre. Billingsley's home Ashwick Grove is north-east of Oakhill, on the route of the Roman Fosseway

the village of Oakhill, which today has the main concentration of houses in the parish, grew up during his lifetime as a result of the establishment of the Oakhill Brewery.

The area is very hilly, the highest parts being well over 250m above sea level, the lowest less than 150m, with many ups and downs between. While not in itself a great altitude, the topography results in the Mendip Hills taking the full force of the prevailing westerlies from the Atlantic, so the climate is noticeably cool, damp and windy.* Ashwick is frequently 'in the clouds'. Billingsley himself said that having come up from the Somerset Levels onto Mendip 'you feel yourself comparatively in Lapland'.[2] This climate naturally has an effect on its agriculture.

Ashwick was small for a parish, both in population and extent: fewer than 700 people then, in an area of about 1,500 acres. Even for those times the communication network in terms of roads was very poor; Mendip was difficult land to traverse. In Billingsley's youth a good deal of the parish land was marsh, scrub and uncultivated waste ground, especially on the hill tops. The underlying rock is mainly limestone. In medieval times the high part of the surrounding land was within the 'Forest of Mendip' so not available for cultivation.† By the mid-eighteenth century this was common land, generally used for grazing. Then, as now, the vast majority of cultivated parish land was permanent pasture, with only a small amount used as arable: wheat, oats and beans then being the main crops. The fields were small and irregular in shape, most being between five and eight acres, seldom more. The parish does not appear to have ever used strip cultivation, the only common land being the open 'downs' on the hills, where parishioners held rights to summer graze as much stock as they could support through the winter.

The main occupation was farming. The majority of male parishioners were agricultural labourers, typical of rural areas then. The predominant stock was sheep, with a small number of cattle, pigs and fowl. Early in Billingsley's life production of wool was paramount: this was the time of the wool trade and this part of Somerset played an important part in it. Top men in the wool trade were clothiers (owners and managers in the cloth industry) and hosiers (stocking makers). The local speciality was knitted stockings, using hand knitters in an outworker system; a large proportion of Ashwick women were employed in that task. The parish lies at the extreme edge of the Somerset coalfield, so many other workers were involved in coal mining and a few in

---

* In the first half of the century the climate was on average cooler and drier than today; in the second half it was cooler, wetter and windier with many violent storms.

† 'Forest' here meaning royal hunting ground; it was largely devoid of trees.

quarrying. Most of the coal output from Ashwick (Moorwood pits) went to Bath.

Near Ashwick (but outside its boundaries) lead, calamine and other ores had been profitably mined on Mendip for centuries. Early in the eighteenth century, apart from those who held land or mining rights, clothiers and to a lesser extent hosiers stood the best chance of doing well financially. Later in the century the local wool trade declined. As this source of wealth and employment waned, the Oakhill Brewery (founded 1767) developed, becoming in its turn the major employer and wealth producer, resulting in the growth of Oakhill village.

Turning very briefly to the national scene, the economic situation changed significantly through Billingsley's lifetime due to a number of different factors, all of which worsened the financial situation. This in turn affected agriculture and the price of land, crops, rents and so on. The 1760s saw a period of bad harvests and rising prices. In the 1770s trade was disrupted by the American boycott of British goods, then by the American Revolution, again affecting prices. Later again the French Revolution and the protracted period of war or threat of war in Europe continued disruption into the next century, together with on-going insecurity, inflation, and social unrest.

Naturally the poor were impacted: during this period there was an huge rise in the local parish poor rate expenditure throughout England – to meet the need the average national poor rate charge per head increased by 26% in the eight years between 1776 and 1785.[3] By 1803, nineteen years later, it had gone up by another 144%. But the increase in poor rates generally was due to more than just disruption of trade and the steep increase in the price of wheat (hence of bread) – for instance, the increase was much affected by the Enclosures. Local issues could also be important. As one example, the poor rates for nearby Shepton Mallet were levied nine times in 1774 due to severe need, eventually totalling £6 18s per head in that year.[4] In this case, another major reason for the serious local situation was that conditions in the wool industry had led to unemployment. For Ashwick itself things were slightly better as the establishment of the brewery had increased the chance of employment, although like Mendip as a whole it was still a relatively poor area.

Land rose in value significantly during this period. It was estimated by Locke that between 1747 (the year of Billingsley's birth) and 1796, the value of land in his district of Somerset increased more than fourfold.*[5] In parallel, rents also increased. In common with many places in England at the

---

* Richard Locke estimated that a farm of 100 acres worth £1,500 in 1747, would be worth £6,750 by 1796.

time the majority of the parish had one major landowner, in this case the Fortescue family of Castle Hill in Devon (first barons, then earls; they owned vast estates in England and Ireland, in addition to other financial interests). The manor of Ashwick had been purchased by the Fortescues in 1671.[6] They continued to acquire more land and property in the district into the early eighteenth century, then sold off most of it by 1803.[7] Fortunately, they had kept good estate records, which have survived. Much of what we know about Ashwick and its inhabitants in that period is due to them. By the mid-1760s the Fortescues had Enclosed land they owned on Ashwick Down.[8] Legally accomplished by an early and unusual form of Enclosure act, it included most of the common land of the parish itself.* The main Mendip common land nearby, however, was not Enclosed until 1782.

The Fortescues made their money locally by leasing land they owned to tenants, who often then leased it on to sub-tenants. Landowners employed local land agents to organise the rentals and collect monies due: for much of Billingsley's time the agent in Ashwick was William Miles of nearby Ston Easton. The system used for a lease was three lives or 99 years, as was then usual in the west country. In addition to details of the land involved, leases give useful information about the local families and their wider connections.

Another extremely important factor at the time was religion – especially in the case of Billingsley's immediate forebears, who were strongly nonconformist. In the late seventeenth century dissenters (as nonconformists were generally known) were still subject to repression. For example, their ministers were prevented from preaching in incorporated towns, making villages and rural areas such as Ashwick likely places for nonconformists to gather. They usually met in fields or private houses: it was safer and more discreet. There is evidence that in Ashwick itself nonconformists met in the woods next to Ashwick Grove, guarded from the law by sympathetic helpers.[9] From the nearest road a valley leads down into the depths of the ancient Harridge Wood, from where it would be easy to disappear.

By the end of the seventeenth century nonconformists had become very numerous in the whole of the south-west, especially in rural areas, Ashwick included.† While the established church was frequented mostly by the 'better sort' (of whom there were very few in Ashwick), the working man was more often a staunch nonconformist, especially in places with occupations such as mining,

---

*   This act was unusual in that it was prompted by the need to complete legal transfers disrupted by the death of the previous earl.

†   At least 14 nonconformist chapels were established in Ashwick between the late seventeenth and mid-nineteenth centuries.

as in Ashwick. But in one sense this parish was something of an exception: the Billingsley family and the various other partners in the Oakhill Brewery (who together made up most of the 'better sort') were strongly nonconformist. Following the Act of Toleration (1688) things eased considerably. Chapels for dissenters were now possible, and they needed ministers.

It was this that had prompted the arrival at Ashwick of the Billingsley family, who were then in need of a home. Billingsley came from a long line of churchmen: going back several generations almost all the men were ordained. Several of his forebears were well known and highly thought of in their own circles, yet often denigrated in others. This must have had a profound influence on his upbringing and moral values, especially as religion was a much more important part of life in the seventeenth and eighteenth centuries than it is today – publicly if not privately. It certainly affected some aspects of his later career. The best-known of his forebears was his great-grandfather, the Rev Nicholas Billingsley senior (1633-1709) who gained a well-earnt reputation for religious poetry and skill at preaching.[10] Although ordained into the anglican church Rev Nicholas senior soon became notorious for his conflicts with the established church. He was twice thrown out of a living for his views and suffered years of on-going verbal abuse. He eventually settled in Bristol. Two of his three sons were also ordained: Richard and another Nicholas, the Rev Nicholas Billingsley junior (1675-1729).

Rev Nicholas junior was John Billingsley's grandfather (see the family tree at the end of this chapter). He was the first member of the Billingsley family to come to Ashwick, where he founded, in Sir Jerom Murch's words, 'the excellent family at Ashwick'.[11] It was a small family, but it became well respected in the district. Ashwick was a good choice: with an existing nonconformist community and lacking the restrictions that a more populous area would have had the group was able to thrive.* Rev Nicholas junior took up the post of nonconformist minister in or soon after 1695. He married Mary James, from a local nonconformist family with many members working in the wool industry. The previous tenant at his new home had been a James.

As the law and attitudes had changed Rev Nicholas junior did not suffer difficulties with church authorities to the same extent as his father, but nonconformism was still an uncomfortable place to be. He was a member of the group that came to be called the presbyterians. It is not known where he originally held meetings, probably at Downside near Shepton Mallet where

---

* The local anglican church (Ashwick St James) was still only a chapel at that time (the mother church being some six miles away at Kilmersdon) and was sited at a distance from most of the Ashwick population. This must have discouraged attendance.

the first nonconformist congregation had already been established about 1688. One part of that congregation moved to a chapel in Ashwick soon after his arrival, with him becoming its long-term minister. He appears to have been a very successful preacher, coming to number 200 or so among his congregation. Ironically, however, he had problems in conforming to the prevailing nonconformist beliefs, so suffered his own trials and tribulations.*[12]

Rev Nicholas junior lived at Ashwick Grove, part of the Fosse House Tenement (see chapter 13). In addition to his work as a minister and some religious writing, he became known nationally in certain circles for giving houseroom there for extended periods to at least two prominent nonconformists who were themselves having trouble because of their beliefs. First, Hubert Stogdon, whose preaching in Devon had led to accusations of arianism and even atheism, 'raising such a clamour that he was obliged to leave the district'.[13] Stogdon stayed at Ashwick Grove for several years. Rev Nicholas junior also gave refuge to the famous divine Dr James Foster (1697-1753) who, according to Rev Collinson, 'secluded from the fury of bigots and the cares of the busy world', withdrew to the summer house to write his religious tracts.[14] Ashwick Grove was therefore very well-known in religious circles. The fame – or notoriety, depending on one's point of view – of these predecessors later affected attitudes to John Billingsley in certain circles, quite apart from anything he himself achieved. The Billingsley name was already widely known.

Other than the religious altercations and some small land transactions little else is known of the Rev Nicholas junior. Either he, or (more likely) one of his predecessors at the Ashwick Grove site, built the first house there. It is not known how he made a living – the contributions of his flock cannot have been great. He is often recorded as 'Mr', so clearly had some substance and standing in the district, albeit not financial.† Rev Nicholas junior died in 1729; his remains, together with other members of his family, now lie in the anglican churchyard at Ashwick St James. A handsome (grade II) railed tomb

---

* His difficulties were prompted by the crisis in nonconformist circles which led to the Salters Hall Synod, following which there was a split in the presbyterian movement. Rev Nicholas junior himself was said to have taken up arian views (denying the consubstantiality of the Trinity) and to have been responsible for establishing arianism in the district. At one stage, he claimed, extreme and unfounded rumours were spread against him; even members of his own congregation then briefly rejected him before accepting his explanation. He felt forced to publish a book containing a rebuttal of the slander, giving great detail as to his religious beliefs and revealing a series of strong disagreements over doctrine, including arguments between himself and other local ministers. He later recovered his excellent reputation.

† 'Mr' was also the style often given to nonconformist ministers at the time.

bearing an inscription dedicated to them says that he and his wife are buried 'near this place' (it is generally assumed that the structure was placed there by his grandson, John Billingsley). Rev Nicholas junior's nephew, Rev Samuel Billingsley (1699-1777) then followed him as nonconformist minister at Ashwick, serving there until 1747 (co-incidentally the year of John Billingsley's birth).

Rev Nicholas junior and his wife Mary had at least four surviving children. There is no record of the baptism of any of them (as nonconformists they were very probably baptised by their father) but there is a record of their birth dates pasted into the register of the parish church of Ashwick St James some years later (as was then required by law – though this was not always honoured). Neither of the two sons were ordained. By then the very strong religious tendencies of the previous generations seem to have been much diluted. The elder of the two, John Billingsley senior (1700-1774), became a clothier in Shepton Mallet. When he started work this would have been a lucrative line of business as the Shepton wool industry was still in its heyday. He does not seem to have undertaken a formal apprenticeship as would have been usual in those days, probably learning the trade from one of his mother's relatives. He married Mary Greening (1713-87), who also came from a nonconformist background. John senior, incidentally, is usually described as being 'of Shepton Mallet' (the adjoining parish to the south) or, as in his will, 'of Stoke Lane' (now Stoke St Michael, the adjoining parish to the east) – never 'of Ashwick'.* This may be because he worked in Shepton, but it is unlikely he lived at Ashwick Grove during most of his adult life. He is almost absent from Ashwick records, apart from being named as a 'life' on some family leases he is not known to have held land in his own right.

The younger son, Nicholas Billingsley (1703-1774), became a hosier (stocking maker). In October 1714 he is recorded as beginning an apprenticeship under his uncle William James, 'hozier of Shepton Mallet' (he would then have been only 11 years old). By 1724 he is described as a 'stockingmaker of Ashwick', when he took on an apprentice of his own; then in 1741 he took on his uncle's son, Charles James, as an apprentice. From the evidence available Nicholas became the more successful brother financially; for example, he was able to buy up tenancies. Following the death of his father in 1729, Nicholas took on part of the lease of Ashwick Grove.[15] He was known as 'of Ashwick', so probably lived at the house. Nicholas continued to accumulate land through

---

* The property was at the junction of three parishes: the main house was in the parish of Ashwick, with other parts of the holding, even the garden, in Shepton Mallet and Stoke Lane.

tenancies, including several farms in Ashwick, where possible buying rather than leasing. By the time of his death he held what Fortescue's agent described as a 'small estate' (see chapter 9).[16]

Although Nicholas was not ordained records give the impression that he may have been rather more interested in religion than his brother. For the son of such a fervent nonconformist it is surprising to find him recorded as a churchwarden at Ashwick St James in 1737, and engaged in church business at many other times. But he is believed to have remained nonconformist in conscience, establishing nonconformist charities in his will, for example. By this point, then, there must have been a rapprochement between members of the established church and nonconformists – at least in Ashwick. Nicholas did not marry. The two brothers died within months of each other in 1774 and are buried together at Ashwick St James. They are said to have been very close throughout their lives, the inscription on their tomb reading 'inseparable in life as in death'. The next generation included John Billingsley.

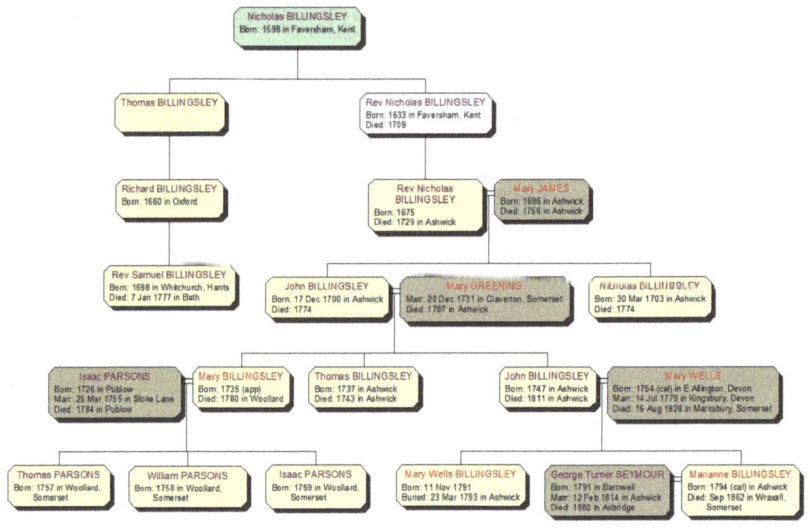

Billingsley Family Tree
showing only those mentioned in the text

2
John Billingsley's Life – a Brief Synopsis*

Despite all that is known of his forebears very little is known about the early life of Billingsley himself. We do not know precisely when or where he was born.† The birthdate of 1747 is usually used, calculated from Billingsley's age at death (64 in 1811). Similarly, his place of birth is usually said to have been Ashwick Grove, though there are other possibilities – perhaps just outside the parish. He was the son of the elder brother, John senior (see the family tree given at the end of the previous chapter).

We know only the bare facts about Billingsley's immediate family. His parents (John senior and Mary Greening) married in 1731 and had three recorded children: Mary, born 1735, and Thomas, born 1737 but died in 1743. After Thomas it was another ten years before their son John was born. His elder sister Mary went on to marry Isaac Parsons of Frome and move to Publow in north-east Somerset in 1755, when Billingsley was only eight years old. Mary and Isaac had nine children. The families seem to have remained close even though the Parsons had moved away.‡

We know nothing as to Billingsley's formal schooling, but he was obviously well educated: he wrote well, and showed skill in numeracy and logic. It is possible (though cannot be confirmed) that he acquired these skills

---

* This chapter contains information on Billingsley's personal life, with a very brief sequence of the events that are covered in more detail in later chapters. In addition, a table of key events is provided for ease of reference. Only items not mentioned later in the book are referenced here, all others where they are discussed further.

† Billingsley has no known birth or baptismal record, although it is probable he was baptised by his father's cousin, Rev Samuel Billingsley, who was still the Ashwick nonconformist minister at the time of his birth.

‡ Three of the Parsons' children figured later in some of Billingsley's land transactions. Thomas and William were both named as executors of Billingsley's will, though William pre-deceased him.

through attending Shepton Mallet's Grammar School, the only school close enough for him to attend daily. He seems to have made little use of the Latin and Greek he would have learnt there, and in which earlier members of his family were fluent. He will have attended the nonconformist chapel in Ashwick with his family, and probably sometimes the Cowl Street chapel in Shepton, his peer group there figuring in his lifelong friendship circle. As an adult he was definitely nonconformist and sincere in his principles. He was involved in the chapel at Ashwick, where he is recorded as a trustee.[1] Miles described him as 'one of the Chiefs of the Presbyterian Brotherhood' in the district.*[2] Unlike his uncle Nicholas, though, he does not seem to have become engaged in parish business at Ashwick St James.

Some time before 1769 (when he was 22) Billingsley had already set up in Shepton Mallet as a clothier. He did not serve a formal apprenticeship, presumably learning the trade either from his father or from his James relatives. It is known from evidence given later to a parliamentary committee that not all workers in the wool trade served a formal apprenticeship during the eighteenth century (although the law still decreed that they should do so).[3] Becoming a clothier would have needed capital, most likely provided by his father. Surprisingly, there is no mention of him in his father's will – so perhaps he had received his portion earlier, when he began as a clothier.† After only a short time in the trade, though, he was looking for pastures new.

The year 1774 would have been a difficult one for the family. Billingsley's uncle Nicholas died in February and his father in June, so that he was now the senior – indeed the only – Billingsley male in the Ashwick branch. He must have already been relatively successful in the wool trade and begun accumulating capital within a fairly short time, this was now added to by an inheritance from his uncle Nicholas. Unfortunately, Nicholas' will was destroyed when the record office at Exeter was bombed in the second world war, so we do not know quite what he left; except that Billingsley was the main beneficiary after charitable bequests (the latter are the only details to survive, recorded by the Charity Commission).[4] The will set up two nonconformist charities, with his 'nephew', named as John Billingsley, charged with managing the funds, which he did for the remainder of his life.

From this period on Billingsley's diverse interests plus huge energy began to show. He took on an increasing number of different projects, in addition

---

* For example, a record of 1808 shows that he had acted as trustee for the chapel for land leased by them at Masbury.

† In view of his temperament and of relationships within the family it is hard to imagine – though of course possible – that there was a rift between them.

to his on-going involvement in agriculture begun as early as the mid-1770s. By his own account, from about 1775 he started using fields near his home at Ashwick to try out his ideas in farming. From then on agriculture, in its various forms, increasingly became his main passion. Alongside this, in 1776 he invested in the Oakhill Brewery as a co-owner. He remained a partner for about 35 years, until shortly before his death. In 1777 the Bath Society was formed (later the Bath and West of England Society). Billingsley became an enthusiastic member at its first general meeting. The society took up a great deal of his time from the very beginning until the end of his life.

At the age of 32 Billingsley married Mary Wells of East Allington, Devon, on 14 July 1779. Mary's father was the anglican vicar of East Allington. Billingsley was the first known member of his family for some generations to have married outside the nonconformist tradition, although it may be that her father also had a tendency in that direction. Mary was related to a branch of the Fortescue family through her mother, Dorothy Fortescue (connections were rather more useful then than now, but we do not know whether that was a factor in the marriage). It was twelve years before a child was recorded: Mary Wells Billingsley was baptised at the Ashwick presbyterian chapel in November 1791. Sadly, she died aged only sixteen months and was buried there in March 1793.

During the early 1780s Billingsley became involved with turnpikes, in particular the Shepton Mallet Turnpike Trust, serving on their committee for many years. At the same time he began to buy up large tracts of land on the Mendip plateau (first in Ubley, then Hazel manor), purchasing Stoke Lane manor (adjoining Ashwick Grove) a few years later. Also in the 1780s Billingsley became ever more involved in the Bath and West Society. He was a sociable man, having many friends and fellow agriculturalists in the society. It was an environment which encouraged his enthusiasm and his ideas prospered. In 1789 he became a freemason, joining the Lodge of Unanimity in the city of Wells together with several of his friends, though it is doubtful whether he continued to be a member for long.[5] Meanwhile, he was continuing to experiment in farming methods, winning prizes and writing papers for the society, describing the processes and outcomes. As time went on his papers became better known and his reputation climbed. He began to specialise in methods for transforming rough ground into land suitable for farming: 'improvement'.

By the beginning of the 1790s he had become involved in parliamentary Enclosures, serving as a commissioner for a large number of Enclosure acts. This continued as a major part of his work for the rest of his life. 1791 was a key year in his personal life, when, in addition to the birth of his first child, he purchased the freehold of Ashwick Grove and set about rebuilding the house – which he transformed into a mansion with gentrified grounds. At the same time he also purchased further land in Ashwick and elsewhere, making him one of the main landholders in the district. He also began to be active on committees for canal building and invested personally in them as well as in coal mining. Plus, he became more interested in water and drainage, working on water meadows and overseeing major projects in draining and Enclosing the Somerset Levels.

The 1790s was the busiest period of his life: his work in agriculture, canals, Enclosures and drainage schemes all being at their peak – and by no means his only interests at the time. Throughout this period Billingsley was acting as steward to Earl Waldegrave's estate, centred on Chewton Mendip. While all went well for the first few years, at some point he fell out with the Waldegrave family. Unfortunately, this resulted in a series of legal cases which dragged on until Billingsley's death, at which time the issue was still unresolved, though his reputation for integrity was restored posthumously.

In 1794 Billingsley's second (and only surviving) child was born, named Marianne. In the same year he was commissioned to write the *General View..* at the behest of the newly established Board of Agriculture. The better-known second edition was published in 1795. In 1796 he purchased land at Green Ore, using it as a model for the process of Enclosure, the results of which he wrote up in a prize-winning essay. In 1797 he was awarded a medal by the Royal Society of Arts for another essay on 'Improving Land Lying Waste'.

Returning to 1794, Billingsley had been proposed in that year, without his consent, as a justice of the peace (JP). This was something that apparently every gentleman of the time aspired to. His appointment was unaccountably blocked, why or by whom unclear.[6] A letter from him shows how offended he was by the whole affair. He discovered his name had been 'expunged' from the list: 'I do not like the Idea of being excluded or that portion of disgrace which was attached to such an act of Exclusion', he wrote.[7] Other than his display of indignation the letter is chiefly interesting in that he says he would in any case have refused to sign the oath, by reason of 'my present religious sentiments', a rare mention of his beliefs.* In 1804 his name was again put forward as a JP, this time his name was even added to the roll. However, he protested that he had never wished for the appointment, and again declined to sign for the same religious reasons.[8]

---

* He could not take the oath necessary for becoming a JP. A certificate had to be provided for all public office holders that the sacrament had been received: as a nonconformist he would refuse the Anglican sacrament.

Towards the end of 1796, notwithstanding the recent extensive renovation work at Ashwick Grove, Billingsley and family moved to Bath. They were soon back at Ashwick Grove, but in 1803 Ashwick Grove was rented out and the family returned to Bath. These periods in Bath may well have been prompted by Billingsley's health – he had suffered from asthma for many years and Bath, of course, had the reputation for cures. In 1804 he became seriously ill and was not expected to survive. But he did, thereafter continuing to carry an enormous work load, making further positive contributions to agriculture and to the Bath and West Society, of which he was now a very senior member. Meanwhile, he continued to invest – 'speculate' – in a number of other concerns.

Apparently believing his health was restored the family returned home to Ashwick Grove in late 1808. But it was not to be. Billingsley died there on 26 September 1811, with much unfinished work in hand – and an unfinished legal case hanging over him. He left a wife, Mary, and an unmarried daughter, Marianne. He was buried at Ashwick St James on 3 October 1811.

All Billingsley's many projects and achievements outlined here are discussed in greater detail in the chapters that follow: the Timeline below indicates the start of each of his main interests.

Timeline of Key Events in Billingsley's Life

Year	Age	Event
1747	---	born Ashwick, parents John Billingsley (1700-1774) and Mary (nee Greening)
1755	8	sister Mary Billingsley's marriage to Isaac Parsons
1756	9	death of grandmother Mary Billingsley (widow of Rev Nicholas Billingsley junior)
1769	22	recorded as a clothier in Shepton Mallet; active in turnpike trusts shortly afterwards
1773	26	purchases logwood mill with John Bowles
1774	27	deaths of father (John senior) and uncle (Nicholas); bequeathed lease of property at Ashwick and elsewhere
1776	29	riots over spinning machinery in Shepton Mallet, his involvement in wool reduces thereafter
1777	30	joins Bath Society at its third meeting (first general meeting); enters into partnership in Oakhill Brewery with James Jordan, together they purchase land at Ubley; death of cousin Rev Samuel Billingsley
1779	32	marriage to Mary Wells of Kingsbridge, Devon

1780	33	most active period on turnpikes; visits London for act for Shepton Mallet Turnpike Trust; wins first of many prizes from Bath and West Society (for carrots)
1783	36	purchases Hazel manor with Thomas Parsons; purchases Stoke Lane manor from Thomas Horner; creates a water meadow
1786	39	agricultural tour of East Anglia with Edmund Rack
1788	41	Wins Bath and West Society's first national ploughing match; begins as Earl Waldegrave's steward
1789	42	joins freemasons Lodge of Unanimity, in Wells
1790's	40's	most productive decade: main period of Enclosure work (between 1789 and death served as commissioner on at least 15 Enclosure acts, one alone covering 18,000 acres); begins work on drainage of the Levels; begins active period on canals
1791	44	acquires freehold of Ashwick Grove and other land from Fortescue; rebuilds house and gentrifies land; work on drainage and Enclosure of King's Sedgemoor begun; daughter Mary Wells Billingsley born (dies 1793)
1792	45	undertakes experiment on sheep feeding; begins acting as steward for Earl Waldegrave's estate at Chewton Mendip
1794	47	*General View..* (1st ed) published by Board of Agriculture; daughter Marianne born; purchases land at Green Ore; name put forward as a JP (and again in 1804)
1795	48	*General View..* (2nd ed)
1796	49	visits Lord Shannon's estate in Ireland; moves to Bath temporarily
1797	50	Royal Society of Arts prize for 'Essay on Improving Land Lying Waste'; investment in Smallcombe Pit
1798	51	paper on the 'Uselessness of Commons to the Poor'
1803	56	leases Ashwick Grove to Miss Lydia White; family moves to Bath; member of supervisory committee of the Bath and West Society; 'Billingsly [sic] and Jordan' act as bankers as well as brewers
1805	58	ceases work for Earl Waldegrave
1806	59	wins Bedfordian Medal for 'Essay on Waste Lands'
1807	60	first court case Billingsley v Earl Waldegrave; litigation continues until after his death
1808	61	family returns to Ashwick Grove from Bath
1810	63	withdraws from brewery
1811	64	dies at Ashwick Grove; buried Ashwick St James

Timeline of Key Events in Billingsley's Life

3
The Wool Trade

BILLINGSLEY'S FIRST EMPLOYMENT was in the wool trade. With his father a clothier and his uncle a hosier this was not unexpected: most boys at that time followed in the family's footsteps. As previously mentioned, he did not serve a formal apprenticeship, probably learning the trade from his father or a relative on his mother's side. He soon became a Shepton Mallet clothier in his own right and may well have taken over his father's business in due course.

At that time there were two predominant centres of woollen cloth making in England: the west of England (Gloucestershire, Somerset and Wiltshire), and the West Riding of Yorkshire. In Billingsley's youth Yorkshire's woollen industry was experiencing rapid growth while that near his home in the west of England was growing much more slowly. By the time he was adult the wool trade in the west was already beginning to experience the slow decline which, with time, would see it give way almost entirely to Yorkshire's, with significant economic and social fall-out. He would certainly have been aware of the change, and with it the increasing challenge to success.

Production of woollen cloth was complicated and the fabric expensive, particularly in the west which made the most superior cloth. In 1757 Josiah Tucker described the industry in the west:

> One Person, with a great stock and large Credit, buys the Wool, pays for the Spinning, Weaving, Milling, Dying, Shearing, Dressing etc. That is, he is the Master of the whole Manufacture from first to last and perhaps imploys a thousand Persons under him. This is the Clothier, whom all the rest are to look on as their Paymaster.[1]

Small wonder the 'Masters' referred to – often called 'gentlemen clothiers' – were in effect merchants who could become very wealthy, while most workers remained poor. Clothiers had a significant amount of money tied up in the process. The workers, however skilled, had no way to accumulate sufficient savings to invest and so better themselves, leading not only to poverty but also discontent. Protests and strikes dogged the trade, with the weavers being the most vociferous, though workers in other sections of the trade were also unruly. In contrast, the system in Yorkshire meant that the clothier there had a much smaller financial investment in the process and the workers had some stake in it (for example weavers owned their own looms), so rather better labour relations prevailed.

Shepton Mallet had long been an important wool town in the region, making it a wealthy area. There are said to have been at least 30 clothing mills in Shepton, all crammed into the short stretch of the river Sheppey that passes through the town.* Yet while there are several histories of the wool trade in the west, despite the size and quality of Shepton's claimed output, remarkably few do more than simply mention the town.[2] Perhaps this is due to events to come.

The area around Shepton specialised in stocking-making. While the best woven cloth needed fine wool, stocking making used short or medium staple wool, so could take on the coarser end of the 'clip'. In the opinion of an experienced Shepton weaver (examined by the parliamentary commission on the woollen trade in 1802), most cloth woven in Shepton was coarse in 1756, but by 1776 it was 'fine rather than coarse' and finer than that produced by nearby Trowbridge, for example.[3] Shepton was also known for its excellence in wool-stapling, the highly skilled process of grading the raw wool. The clothier often took on this work personally; Billingsley himself became very skilled in it, later frequently serving as a judge in competitions.

In his early days the Mendip sheep, a hardy upland breed producing high quality, curly, short staple wool, was still the most common in the immediate district. It was much favoured by Billingsley. Youatt describes it as 'a very peculiar and valuable breed inhabiting the Mendip Hills', while Billingsley says it was 'of a sort that will thrive on the poorest soil, and fatten on such land as will scarcely keep other sorts alive'.[4,5] There seem to be no existing images of the Mendip sheep, but it is said to have been intermediate in its characteristics between the Exmoor and Dorset, with small horns. However, it was so commonly crossed with other breeds in order to produce a larger and heavier sheep for a better financial reward that the original breed is now extinct. The briefly fashionable merino took its place during Billingsley's lifetime as provider of the best wool in the area, but was itself soon supplanted.

---

* Very few of these sites can be found today and little is known of the businesses and wool transactions that must have taken place. However, there are records of many individual clothiers operating at the time and a number of their grand houses remain. Interestingly, a large proportion of Shepton clothiers were nonconformist.

By the time Billingsley entered the trade his father and uncle had been well established in it for over 30 years, although little is known of their operations. There is no record of a mill related to the family. As John senior was so close to his brother Nicholas, it is likely that they worked in concert, as so often in families at the time. For example, John senior may well have provided Nicholas with the yarn for stocking making, his wool-stapling methods being a crucial part. We do not know where in Shepton John senior operated, but Nicholas worked mainly in Ashwick where there were many stocking knitters. Both John senior and Nicholas must have done relatively well financially, but neither became really wealthy.

Unfortunately, there is very little information available as to Billingsley's own activities in the wool trade. In 1769, when he would have been 22 and his father 69, one of them took on a William Jenkins as an apprentice.[6] In view of his father's age it was very probably John junior. The apprentice would have helped his master achieve greater output in his work. Billingsley may then have been working with his father or on his own account: if the latter he would have needed some financial reserves to set himself up. Not only did the clothier need money to acquire premises – mills were in short supply and the river crowded – he also needed money to buy the wool and to pay the various workers as the wool went through the process. The money was tied up until the very end, when, on selling the finished cloth he hoped to get the best return in a volatile market. So he might hold onto his cloth for a time until the price was right. Some clothiers operated as factors, mainly buying and selling the finished cloth rather than supervising the whole production. It has been suggested that Billingsley (and possibly his father before him) worked at least partially as factors.

Billingsley must have been making sufficient profit from his early endeavours to allow him to undertake a major purchase outside his usual work. In 1773 he went into partnership with John Bowles, a friend and clothier of Coleford.[7] They are described as 'Drysalters': that is dealers in dyes which would have been used for woollen cloth. Together they purchased a 'new erected dyestuff mill at Stoke Bottom', on the small stream in Stoke St Michael parish, adjoining Ashwick. An existing site had been recorded there in 1760. They would have used the mill to chop imported logwood (also called bloodwood) then steep it to release a natural dye, using machinery driven by water. The dye could be used to make various colours depending on pH; the one most used in the wool trade being a strong purple. So even at this early stage Billingsley had diversified, seizing a chance that became available. But the project did not last long: in 1783 Billingsley bought the

Manor of Stoke Lane in which the logmill was situated, but he sold the mill in 1784. Beyond that 1773 purchase no other information can be found tying Billingsley to the wool trade until 1776, when he would have been 29, and the most interesting evidence of his time as a clothier appears, described below.

Throughout the period there had been a good deal of unrest in Shepton. The town's workers were regarded as particularly unruly even for those times. During the previous fifty years they had rioted many times for various reasons: they rioted over the price of corn and of bread, as well as over wages and conditions of labour. As a result soldiers were often stationed in the town. In 1776 there was another riot, this one related to the wool trade. Rumours were rife that machinery was being introduced which would lead to loss of work and wages. And wages were already depressed, resulting in poverty. As evidence of this, in 1774 the land agent Miles had described Shepton as a parish where 'the Poor are very numerous and burthensome', adding that the poor rates collected from property and business owners were very high (high poor rates were in fact typical of industrial areas).[8]

In late spring 1776 the weavers, hearing of a now definite plan to introduce spinning machinery into the town, circulated a leaflet, stating that use of the projected spinning machine, or 'Spinning Jenny', meant that two persons could do the work of eight spinners, which would deprive six out of eight of work. This would put them 'on the Parish' (supported by the proceeds of the poor rate), so the leaflet asked for subscriptions for their relief. In a letter to the newspaper the 'gentlemen Clothiers' disputed the leaflet's facts, declaring that the spinners would not be put out of work – the machines would in fact tend to increase the amount of work available.[9] While the machinery may 'inconvenience' the spinners it would nevertheless lead to more employment, for the clothiers had recently been unable to fulfil orders due to lack of yarn.

Undeterred by the reaction of the spinners a group of clothiers went ahead with their plans. This group included Billingsley – given his entrepreneurial spirit and subsequent events he was very probably a leader in it. Apparently, 'some persons from the North, skilld in the use of these machines' were brought to Shepton to instruct the workers in how to use the new spinning jenny. A machine was installed in the workhouse. The Shepton workers objected and stopped work. Men from other places joined them, marching to the town. The Shepton clothiers described the events in a letter of 20 May 1776, sent to the War Office:

a mob of Fifteen hundred and upwards, many of whom were armed, went to the Workhouse…and after throwing large stones at the Door etc over the walls of the court they violently broke open the Door wch was locked and prostrated the wall of the court threatening to destroy the whole Work House, set fire to the Town and take signal vengeance on all that should resist them…after a long time and with utmost difficulty they were prevailed upon to desist for that evening.[10]

The clothiers felt they had 'no power or resources to quell this disturbance or defend ourselves – so earnestly desire that you would send soldiers to come to our immediate assistance'. The letter was signed by eleven clothiers of the town, including Billingsley. Five days later a further letter was sent, certifying that the above violence had taken place as stated, and stressing the need for a military force, signed by the three Shepton magistrates. When received by the War Office this letter was endorsed with the words: 'a Troop or Detachment equal to a Troop from the Royal Dragoons, mounted or dismounted, to march without loss of time from Dorchester to Shepton Mallet'.*

But the problem was not solved by the arrival of soldiers. In late June 1776 something triggered another more serious riot, the result of this one being that Shepton Mallet has the unfortunate distinction of being the first town in England to suffer riots related to the cloth trade involving actual machine breaking. However, the Shepton men may not have been the prime instigators as they were definitely aided and abetted by workers from other wool towns. The *Norfolk Chronicle* reported receipt of a letter from Shepton, dated 2 July 1776, saying:

> Last Monday in the evening a riotous mob of weavers, shearmen, etc, collected from the towns of Warminster, Frome etc assembled together, and proceeded to the town of Shepton Mallet with intent to destroy, under cover of night, a machine lately erected by the clothiers, for the advancement and benefit of the manufactory, and to pull down the houses, and take away the lives of those persons who encouraged and promoted the use of it.[11]

The clothiers had notice of their coming, the dragoons were summoned, but two of the three magistrates called out to help had gone away again by the time the mob appeared in the town. Meanwhile, being temporarily unhindered, the rioters broke into the poor house, destroyed the new machine and caused other damage. At which point:

---

* A troop of dragoons is believed to have then numbered about 60 men.

the military, preceded by the remaining Magistrate, advanced and secured five of the ringleaders, but in conveying them to the prison, they were attacked by the whole body with an intention to effect a rescue...the proclamation [Riot Act] was read by John Strode Esq, who very humanely advanced to the mob, accompanied by a principal clothier, and endeavoured...[to] persuade them to disperse.

However, the rioters refused to leave the town unless those captured were released. They began to 'most cruelly stone' the soldiers, who were then ordered to discharge two rounds above their heads. The attack continued with 'redoubled vigour', so the command was given to fire. According to the newspapers one fell and six were wounded.* Chastened, the mob dispersed. The coroner recorded a verdict of 'Accidental death by the military, under the command of the civil power'.

No Shepton burial record in the few days after the riot mentions death from that event until on 12 July, when James Helps was buried, 'that was killed in the riot about the Spinning Jenny by a dragoon'. And on 24 July Joseph Harden was buried 'that was wounded by the dragoons about ye Spinning Jenny'.† Two weeks after the riot a notice appeared in the *Bath Chronicle*:

Various reports having been propagated through the country, prejudicial to the character of Mr John Billingsley, Clothier of Shepton Mallet, by representing him as the principal cause of the late disturbance in Shepton, and as the sole introducer of a machine lately erected for the promotion and benefit of the Manufactory – We whose names are hereunto subscribed, do certify, that such reports are entirely void of truth, and that the said Machines were sent for and introduced by Us unanimously as a Body, by and with the consent and approbation of the Weavers themselves; the said Mr Billingsley being no otherwise concerned than as an individual. And we do also declare it to be our opinion, that his conduct in the late affray was irreproachable, and perfectly consistent with the nicest principles of Humanity.[13]

---

* Samuel Curwen, an American Loyalist visiting Shepton in September 1776, claimed that 'three were killed and a number wounded, but for Government orders to avoid bloodshed more would have been'.[12]
† Helps appears to have come from Beckington rather than Shepton. It is not clear where Harden came from. The dates of the burial records suggest that Helps died at the scene (or immediately afterwards) and Harden died some days later, from his wounds.

This was signed by eleven clothiers from the town, the majority being relatively young men who were Billingsley's peer group (and who were also, from later evidence, among the most progressive and keen to modernise). From the wording of the above notice we gather that it was John Billingsley himself who was the clothier who had faced the mob. It was a confident action on his part, apparently diplomatically undertaken, foreshadowing traits more obvious later in his life. It is also interesting that the notice claims the machines were introduced 'with the consent and approbation of the Weavers themselves'. Could this mean that the protesters from Warminster and Frome were actually the agitators, and the weavers of Shepton less resistant to accepting machinery? We will probably never know.

The action of the Shepton clothiers in introducing machinery was supported by the wider trade in the west. At the beginning of September the newspapers reported that:

> a respectable and numerous meeting of the Clothiers at Bristol gave thanks to the Clothiers of Shepton Mallet for their spirited conduct in introducing a new Spinning Machine, which was conceived to be of general utility to the manufacturers of this kingdom…[14]

The notice reporting this motion was signed by 96 clothiers, several of these being companies or partnerships, indicating the extent and size of the trade at that time.

But for those in Shepton the threat had not gone away. On 10 September the clothiers wrote again to the war office, that there 'having already been two dangerous riots here…the last would have been very fatal without the presence of Military, only now preventing the Populace from committing further outrage and violence' and requesting that *two* troops of dragoons should be stationed in the town over the winter. This was signed by 15 clothiers, not including Billingsley.[15]

A month later another general meeting of the clothiers of the west of England, this time in Bath, unanimously resolved to continue to adopt spinning machinery and encourage its use, as

> unless the use of such Machines be adopted, the Woollen Trade in the West of England will daily decline, and … the Manufacturers of the North of this Kingdom by the general use of these Machines, will Enjoy such great advantages as cannot fail to secure them the preference at all markets there.[16]

There were no more riots in the short term. But the damage was done: the refusal of Shepton's cloth workers to accept machinery, like most in the west, meant that the trade moved north. The reasons behind this refusal have been much debated – to what extent were they simply objections to machinery *per se*, or to change of status, and to what extent were they caused by poverty and distress, it being the days before labour unions of any kind? Whatever the reasons, the riot affected Shepton profoundly: its refusal to accept machinery meant it never regained its place in the trade and continued to suffer a gradual decline. Some say its loss of wealth and status, due to the failure of the wool trade, is visible in the town even today. From then on fewer young men went into the wool trade in the west, particularly in Shepton. Noticeably, several of Billingsley's peers who had signed the letter in support of him, soon moved into different areas of work.

The next year (1777) a John Cook from near Chard wrote to the newly established Bath and West Society about the performance of the spinning jenny. He felt it was much in need of improvement, prompting them to offer a prize for a better design. The prize was never taken up. In any case clothiers may have felt the spinning jenny would never be effective enough to have much impact on wool production.[17]

Machinery was introduced to the trade in the west, but more slowly than many clothiers would have liked. Billingsley himself was unequivocal in his view of the situation: in the *General View..*, written in the mid-1790s when he had already left the industry some time previously, he says:

> Machinery must and will be universally introduced otherwise the districts, where it is not used, must be sacrificed to those where it is…the progress of machinery in the west is rapid: its adoption there in a short time must become universal on a principle of self-defence; without it, the trade must migrate from that district to the north.[18]

Although by 1803 it was reported that there was no longer any objection by the workers to the introduction of machinery, by then the Shepton wool industry had already declined.[19]

After the riots Billingsley himself very soon moved out of the wool trade. He was an astute businessman and must have quickly seen that by that time wool was not the way to prosper. Even so, it is rather surprising that in March 1776, two months *before* the date of the first riot, he had already taken on a partnership within the Oakhill Brewery. Maybe he had foreseen the future. Early evidence of his skill as a speculator.

4
Turnpikes

Billingsley's first involvement in matters outside wool was related to roads. Throughout his working life he put great emphasis on the need for good roads: 'Amend the public roads... nothing so much contributes to the improvement of a county as good roads; before the establishment of turnpikes, many parts of this county were scarcely accessible'.[1]

In his early years defective roads were the norm in Somerset – and in England as a whole. Roads were particularly bad on Mendip. Substantial improvement did occur during his lifetime, largely due to the establishment of turnpike trusts, in which he played his part locally. Prior to the turnpike era roads had little in terms of foundation or surfacing and wheels were usually narrow, so most developed deep ruts, complete with either thick mud (which could be waist deep) or the whole area blanketed with dense clouds of dust. Arthur Young railed against one with: 'this infernal road was most execrably vile with ruts four feet deep'.* The Highways Act, from as far back as 1555, established road upkeep and repair as the responsibility of the parishes through which the roads ran. This act was still in force in Billingsley's time: each parish was required to appoint a surveyor of highways, who was to organise statutory labour and materials for upkeep of the roads as necessary.

Turnpikes began in the 1600s and rapidly increased in number and extent.[2] In principle it was a simple system. A bar was erected across the road; travellers were charged a toll to pass, in exchange for travel on a better road. Each area formed a trust which undertook to look after their roads. Investors became trustees who oversaw both repair and maintenance of existing major roads and in necessary cases built new ones. The funds generated by tolls were

---

* In addition, the state of the roads meant that travel was inordinately slow. It took four to four and a half hours to travel on horseback between Ashwick and Bath (sixteen miles), and three to four hours to Frome (ten miles). It could also be expensive: in 1775 a public coach from Oxford to Bath, a distance of 62 miles, took from 5am to 8pm and cost a total of £4 8s [Woodforde].

split between costs of maintenance and return for investors. Between 1750 and 1770 the number of turnpike trusts trebled, transforming the major road system within a very short time.

Each trust had to have an initial act of parliament, but as acts were in force for a maximum of 21 years most were followed by further acts, with associated cost and effort. Any significant increase or change to the road, purchase of land or toll building would also often require yet another change in law. Trusts were not-for-profit and maximum charges were set by law. Each trust might manage quite small stretches of road, so there could be several individual trusts in one area. In Billingsley's area of Somerset there were nine separate trusts (see the plan below).

Apart from the acts themselves the government provided very little assistance for trusts overall; there was no central or even county-wide finance for transport. The entire turnpike system arose through local enterprise and was managed and financed locally. There were many weaknesses – and many critics – of the system, especially at the start, but as more turnpikes were built the main road network definitely improved. Turnpike roads were maintained to a much higher standard than others. In addition, provision of gates and tollgatherers, tollhouses and milestones also had to be provided and managed by the trust. Meanwhile, in rural areas, such as Mendip, local and minor roads not subject to tolls continued in the old tradition: parish funding for maintenance with statutory labour. Parishes were frequently in default, meaning the majority of roads not turnpiked remained in poor condition throughout the eighteenth century. Billingsley was active in trying to ensure Ashwick's local (non-turnpiked) roads were maintained.*

The proliferation of trusts and the morass of different methods and regulations resulted in the need for the first general Turnpike Roads Act, approved by Parliament in 1773.[3] This act made the relevant parish and trust jointly accountable for the state of the roads – otherwise the parish would find itself at the wrong end of a fine – thus it encouraged parish as well as trust engagement. Finance was an on-going issue: trusts often contracted debts and investors became increasingly difficult to find (it was a time of inflation and by law the interest rates were kept unattractively low). These national problems were reflected in the local scene.

In Billingsley's area of Somerset the first trust to be established was for Bath (1707), followed by Bristol (1726). Wells and Shepton Mallet trusts

---

* Ashwick parishioners were typical in their laxity: for example, between 1771 and 1774 there were six summons issued to its inhabitants for failing to perform their statutory duty in repairing the local roads [McGarvie, 1997].

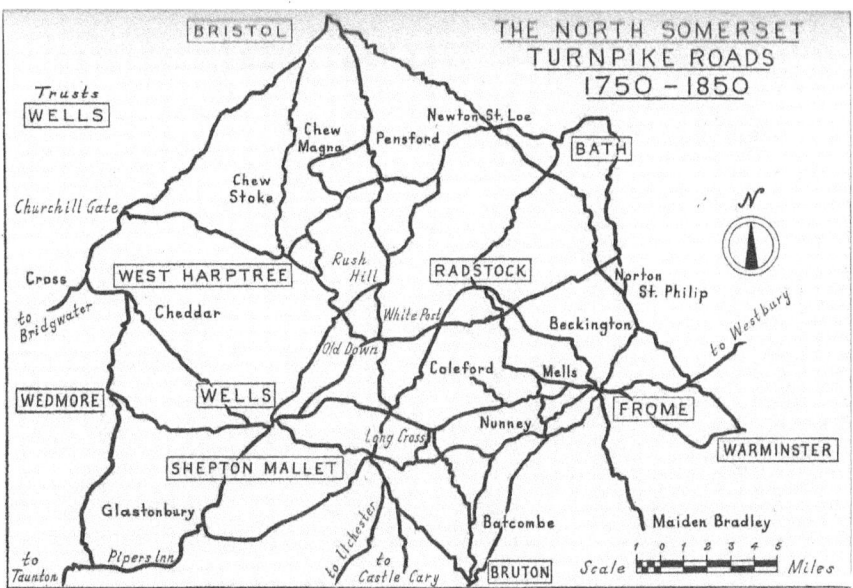

North Somerset Trusts 1750-1850 (Atthill)[5]
Ashwick lies at the centre of this map, between the words Old Down and Long Cross

followed rather later, both being established by acts in 1753.[4] Billingsley's father and uncle were founding trustees for the Shepton Mallet Trust. Billingsley himself was only six when it was formed, but from the early 1770s he was active within it (he may also have invested in one or two of the other trusts, possibly Bath or Wells, but no evidence of this has been found).

Although the Wells trust maintained roads running right up to the border of Ashwick, the Shepton trust was not only his nearest but also by far the most important for Billingsley himself, for several reasons. For example, he made a financial investment in it; he played a personal part in administering the trust and was active as a trustee and commissioner; and two of its turnpiked roads ran through Ashwick itself: the Bristol Rd and the Bath Rd, the latter running right past the entrance to Ashwick Grove.* As was usual, further acts followed for the Shepton trust, another three in Billingsley's lifetime, including one in 1780 in which he was an important player. Involvement in the trust was one of his first forays into public life, while he was still a clothier in Shepton.

The first Shepton act listed the Shepton roads that could be turnpiked and also specified the tolls allowed and the costs. These were still in force for much of Billingsley's day:

---

* The roman Fosseway ran within yards of Billingsley's home, but was so decayed by his time that it was of no benefit to travellers. No attempt was made to turnpike it.

every horse or other animal drawing a carriage or other vehicle, to pay 6d
every horse or other animal drawing as above in pairs, 4d
every horse or other animal not drawing, 2d
every drove of cattle, 10p per score
every drove of pigs or sheep/lambs 5d per score.[6]

The money was to be vested in trustees for keeping and improving the roads, first deducting one moiety (half) of the tolls to pay for expenses in passing the act. Not less than half of the remaining income was to be spent on repairing and improving the roads. Details of when and where meetings of trustees should take place were stipulated. Five or more trustees were necessary for a meeting to be quorate. Trustees were obliged to swear an oath to ensure that they were men of sufficient substance: the requirements being £40 annually in rents and property to the value of £1,000.

The second Shepton act (1765) was a minor one, chiefly designed to give extra years so that funds could be used to repair and improve the roads, and for enlarging the terms and powers of the first act. No names are given for any additional trustees.[7] Billingsley, only eighteen in 1765, must have soon become a trustee: by the early 1770s he already had sufficient capital to qualify. It was probably his first experience in civic life. He continued to be involved in the Shepton trust over many years; his work in it may have been intense for short periods of time (mainly spread over the first fifteen years of his trust membership), but it would never have been a major part of his activity. Trustees met in a public house in Shepton, convenient for Billingsley's work at that time as a clothier. Although no road plans still survive for this trust, fortunately the minute books for trustee meetings after 1776 have been preserved.[8] The trust was then considering petitioning for a further act, part of which would concern the Bath road which needed rerouting between Shepton and the White Post, via Nettlebridge, a notoriously difficult stretch involving a steep valley. This road went right past Ashwick Grove and improvements to it would also have affected transport for the Oakhill Brewery, which had significant transport needs. In the process of acquiring the land necessary for the change of route Billingsley was asked to write to the owner, Lord Fortescue.

Shepton Mallett Dec 17, 1779

Sir
I am desired by the Commissioners of our Turnpike to inform you that they propose applying to Parliament this Session for an extension of their Act & if

they succeed, it will be necessary in order to grade an Hill of very steep Ascent to pass thro' a few Fields of his Lordships Land situate at Nettlebridge in the Parish of Ashwick --- As this Intrusion to his Lordships Property cannot be attended with any material Inconvenience rather to the Tenant we presume his Lordship will grant us his Patronage & support as the Road will be of great public utility. We have also desired Mr Miles to write on this subject.[9]

In January 1780 it was 'resolved and ordered that William Paget and John Billingsley should be a Committee to attend and wait on the House of Commons with a Petition now prepared and agreed in order to carry into execution the several matters contained therein'. The treasurer was to supply them with the funds necessary. For Billingsley, who is not known to have travelled far from home prior to this, it would have been a rare visit to the capital. The act was passed by parliament in spring 1780.[10] In May 1780 the Trust resolved that a group of nine, including Billingsley, should be appointed a committee 'to take such lands as will be necessary to divert the road at Nettlebridge Hill and to widen and alter and form this road'. Later in May, Joseph House (who lived close by) qualified as a trustee and was immediately elected to superintend the section of road in question. He was paid several instalments of £50 for his work and expenditure.

Recognising the need to raise additional finance to cover this ambitious programme the trust began to issue bonds in small numbers, all taken by existing trustees but without, it seems, great enthusiasm. Billingsley was the last trustee to take up a 'Deed Pole' for £50 (interest at 4.5%) so that in total the trust raised £7,500 by this method. At the same meeting they resolved to borrow a further £950 (also at 4.5%). Accounts for the trust at this time seem to have just balanced. By 1783 money was being 'laid out' to the proprietors of land needed for the rerouting of the road at Nettlebridge. There are no details of the line the road would then take, but a later map (below) shows where it must have gone.

Meanwhile other aspects of trust business were also occupying the trustees. They needed to appoint tollgatherers. Changes of tollgatherer were frequent: between 1780 and 1792, for example, four different men succeeded one another at the Nettlebridge Gate. Honesty will have been a prime consideration: tollgatherers were required to swear an oath each time they deposited toll money. Those chosen by the Shepton trust appear to have abided by their oath; no irregularities in returns are recorded. However, as might be expected, honesty and good behaviour were not universal among the local population: a large quantity of pavement stone due for the new road was stolen from the quarry. And 'Edward Lambert assaulted Joseph Langman the tollgatherer at nearby Marchants Hill and broke the locks on the gate and the tollhouse'. He acknowledged the offence and agreed to pay expenses.*

Arranging housing for tollgatherers was another issue that did not go smoothly. Joseph House had been asked to provide a tollhouse at Nettlebridge in 1780, but in 1783 a committee of three, including Billingsley, was appointed to fix the place for a house and 'carry it into execution'. By 1785 there was a new superintendent for the road. In 1801 the tollhouse was moved out of the parish to Stratton Common (presumably to take in tolls from those arriving by side roads). The committee were then to arrange for the land of the old tollhouse to be returned to its previous owner and to pay Joseph House £25 for building the new one at Stratton.

Perhaps this type of activity became tedious to Billingsley, as once the main task of arranging alterations to the road was finished he attended meetings only intermittently, then not at all for three years. Indeed, many trustee meetings failed to be quorate so could not proceed. The committee met only to make resolutions, most of the work being done outside meetings. However, by 1790 the trustees were considering yet another new act. They appointed a new set of nine commissioners, not including Billingsley (he was probably too busy, this period being the beginning of his peak workload). The outcome meant the line of the road at Nettlebridge was again altered. By this stage Billingsley's interests had definitely moved on. He was much less involved in the trust, only taking a brief active role again much later when matters affected the immediate vicinity of Ashwick Grove. Unfortunately, the route chosen by the commissioners in 1791 still did not prove to be a workable solution, and the road had to be relocated yet again in 1830.

From 1800 tolls were let annually at public auction. This was part of the notorious system of toll 'farming', in which someone could buy the proceeds of a gate for a year and realise a substantial sum. The toll at Stratton Gate was 'put up' at £85 2s.1d and 'taken' for £110 by William Stallard. Usually, gates were taken by local men, but in 1801 Sylvanus Hanley, who was not a local man, took several gates, including Stratton. Rumours began. Evidently Hanley was intending to charge double the amount permitted by law, so after a brief fracas during which the trustees threatened him with legal action they refused to put him in possession of the gate and returned his deposit, deducting £37 for costs accrued. No more was heard from Hanley.

In 1800 the trust was managing eight roads, covering 51 miles with

---

* He was lucky that no further action was taken against him, it being usual to refer such miscreants to quarter sessions.

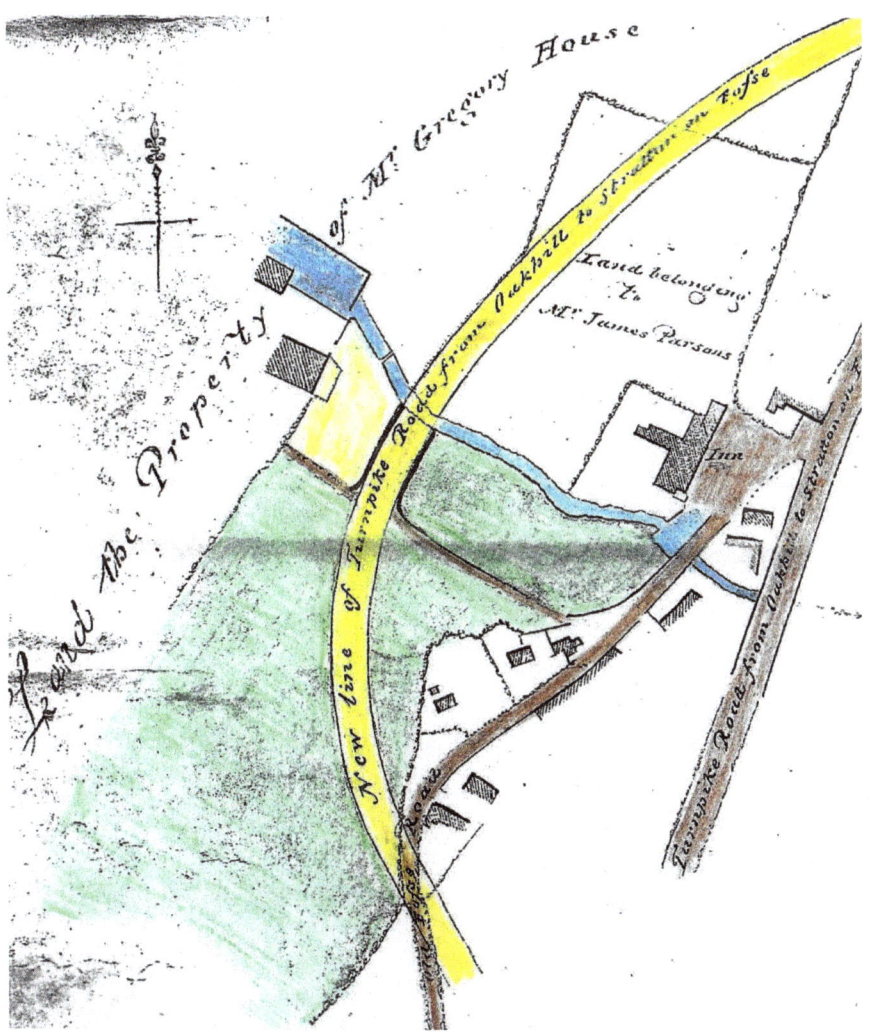

Turnpike Route through Nettlebridge
plan showing the old Turnpike route established 1780 (brown route), the new route of 1791 (grey route) and the proposed new route of 1830 (yellow route) [11]

17 toll gates (a similar number of miles to the Bath trust), with much less difficulty than previously. But by June 1802 the treasurer reported that 'the Trust is considerably in Debt that Bills for Halling [*sic*] stones to a large amount are unpaid and that the Trust Fund does not appear to be sufficient to meet expenditure'. This was a common occurrence at the time. It was also a time of increasing financial crisis in the country, with strong competition for investment. Billingsley did not himself take up further stock in the trust, nor did he take any of the gates at auction, evidently not seeing it as a good

investment. His new investments at the time were centred around canals and land, and as he was now living in Bath it would in any case have been inconvenient to attend meetings at Shepton. But he did remain a trustee.

In 1807, while he was still at Bath, another alteration needed to the line of the road near the village meant a further purchase of land was on the agenda. Despite the trust's financial position, the four commissioners appointed to carry this through (not including Billingsley) decided that it was 'absolutely necessary for the safety and accommodation of travellers to deal with this section of road'. It included the section immediately outside Ashwick Grove, so Billingsley made several of his otherwise now rare appearances at trustee meetings. His neighbour John Tapp, who owned some of the affected land, expected to be paid £30 per acre, but the trustees felt that £20 should be sufficient. At the next meeting Tapp qualified as a trustee – no doubt feeling he would thereby have more say over future decisions.

The trust arranged a surprising number of changes to the Ashwick roads around this time: the point where the Bristol road coming out of Shepton diverged from the Bath road was moved northwards; the line of the Bristol road was moved westwards; the position of the Bath road at the entrance to the village of Oakhill and the continuation of that road going north from there to beyond the entrance to Ashwick Grove were all also moved westwards.[12] These were significant changes, involving new sections of road, not just repair and maintenance – and this for just the small length of roads that ran within Ashwick parish. Mendip, including Ashwick, is hilly country: while these changes were not untypical the extent of them is impressive, especially in the days when all was done almost without mechanical help so requiring considerable manual labour. It makes some of the deficiencies in trust finances seem more understandable.

In 1809, when Billingsley had again stopped attending meetings, his name was put on a committee for changes to the road at Pye Hill, not immediately local. There is no record of his actual involvement. Perhaps the trust was struggling to find qualified – and willing – trustees to act. In September 1811 his name was also put on groups to inspect contracts – but even had he intended to do so he would have died before he could have acted. No more major changes were made to the Shepton trust roads during his lifetime after those allowed by the previous act.

Billingsley's personal involvement in the Shepton trust, and his discussion of the subject in the *General View..*, both give evidence as to his attitude to road improvement. He is of course concerned in his discussion with all Somerset, no doubt shaped by his local experience. Considering, first, private roads (that is those not turnpiked), Billingsley clearly regards the system of statutory labour to be a problem. He expresses impatience with anyone he sees as indolent: those who are expected to do this duty are 'in a state of torpor', 'cannot be roused by the surveyor', and the work done is 'about half what it ought to be'. And the farmer called to statute duty 'goes to it with reluctance, as a legal burthen, from which he derives no benefit'. The remedy Billingsley suggests is for the surveyor to receive the money in 'highway tax': he could then employ such workmen as would do him justice – or if they were indolent he could dismiss them. He does not specify how the highway tax should be raised.

Turnpiked roads were obviously far superior: by the date of the *General View.* he called roads in the northern parts of Somerset 'pretty good', considering the traffic on them, while south of Mendip, the turnpike roads that led to Bridgwater were 'as smooth as a gravel walk'; with the necessary stone splitting being done by men sitting on small sledges, reducing the stones to the size of a pigeon's egg at the cost of sixpence a ton.[13] Initially, he says, there were objections to the turnpikes, for:

> The money collected at the gates was considered as a burthen and the public were for some time loaded with an extra charge for carriage. This, however, did not last for long, for in the course of a few years, a diminution in the price of carriage universally took place, and it has gradually fallen from that time to this. Before the turnpike roads were established, coal was carried on horses backs to the distance of fifteen or twenty miles from the collieries; each horse carried about two hundred and a half weight. Now one horse with a light cart, will draw ten hundred weight, or four times more than the horse could carry; can an insignificant toll be put in competition with this saving?

The speed and comfort of travel generally had also improved enormously. There were far fewer jolting pot-holes and much less damage to vehicles. Between 1750 and 1800 average journey times improved nationally from 2.6 to 6.2mph as a result of better roads. This meant lower transport costs and also decreased wear and tear.[14] After the roads were turnpiked travel became much less expensive and faster than previously. In a letter of 1802, Billingsley gives the expected cost of travel by coach from London to his home at Ashwick Grove (for a prospective employee), as taking from 'half-past four in the morning [arriving] the same evening about 10'. The cost of that journey would come in at just under two weeks wages: £1 11s 6d, when the wages offered were 2 guineas a month (to a highly skilled worker, see note in chapter 13).

But most important of all for Billingsley was the positive effect of turnpikes on conditions for trade. Through his involvement in the brewery in particular he was well aware of the difficulties of transport. He described the 'immense saving of labor' from turnpikes, though the 'establishment of such roads was as unpopular, and the probable benefit as little credited, as those of canals are now.'[15]

5
THE OAKHILL BREWERY

THE OAKHILL BREWERY was established in 1767 by James Jordan and Richard Perkins.[1] It soon became an extremely successful enterprise with a national reputation, lasting for just over 150 years. Unfortunately, the brewery suffered a catastrophic fire in 1925, from which it never recovered. All the company records were destroyed in the fire, so we know very little about its early years.

Although often claimed to be a founder of the brewery, Billingsley did not actually become a partner until 1776, about the time he moved out of wool.[2] This was nearly nine years after its establishment. The three families – Billingsleys, Jordans and Perkins – would have been well-known to each other, living near to one another and all being nonconformists of a certain standing; a close social circle. The young men were of a similar age and at least later, if not already, they were friends. From its foundation Billingsley would have been well aware of the progress of the enterprise and possibly involved informally throughout.

Neither of the two founders had any family or personal history in brewing and neither man served an apprenticeship in it. Presumably, then, they had no prior knowledge of the business. James Jordan (ca 1745-1830) was from a Shepton Mallet family, his father being a clothier. When the latter died in 1764 he left his son a good inheritance for investment in a project. The money was essential, as brewing on any scale involved a large initial capital outlay. Jordan decided on brewing relatively quickly: he was still only 22 when he co-founded the brewery, which became his life's work. In June 1774 he married the other co-founder Richard Perkins' sister Joanna – the two men are sometimes described as 'brothers by marriage'. While others came and went, Jordan seems to have been the mainstay of the brewery for its first 35 years.

Richard Perkins (ca 1750-1825) was from an Ashwick family, well-known in the district for some generations. His father, Dr Joseph Perkins, was the surgeon of Ashwick as well as owning and renting land in the area. Richard junior also began his working career as a surgeon, entering into partnership with Jordan while still practising medicine. By 1774 he had evidently tired of it, advertising in the newspaper that he 'declined Practice in Surgery, except in cases of emergency as an Assistant Surgeon'.[3] This must have given him freedom to spend more time on other things – though these do not seem to have included the brewery. He had an active mind: while the process of brewing may have preoccupied him initially, by the start of the 1770s he had begun to dabble in other areas of science and technology, especially geology. In 1776 he pulled out of the partnership.*

Returning to the establishment of the brewery, the new partners first needed to consider the brewing process. Early in the eighteenth century it was still being carried on in much the same fashion as in the past several hundred years, usually in relatively small quantities, the brewers using the methods their forefathers had used. Brewing was an art. Around the middle of the century they began to recognize that in order to ensure a reliable outcome they needed to take better control of the process. It required expert timing and accurate measurements of heat, moisture and alcohol content.[4] The new approach reflected incoming scientific methods. However, the exact measurements needed were not possible until about 1760, when the thermometer was introduced to brewing, and in 1770 the hydrometer (or saccharometer) – the latter measuring the specific gravity of the liquid, enabling prediction of the percentage of alcohol being created. Together these two instruments transformed the brewing industry from an art to a science. To stress the point, Jordan and Perkins decided to begin a brewery at just the time of this transformation: an exciting new prospect was emerging.

By the late 1760s there were also a number of other changes that made starting a good-sized brewery an attractive proposition – the Oakhill Brewery was not to be the small local affair that had produced most village beer hitherto.[5] First, the government had introduced new tax rules which encouraged a move to larger breweries. Second, the rising popularity of the new 'intire butt' beer, known as porter, which had a better outcome when brewed in larger vessels, led to a move to larger quantities. And not only was porter attractive to customers through its taste, it also kept longer and was

---

* Perkins had a fascinating career thereafter, with interests in geology (he was a friend of William 'Strata' Smith), canals and coal, leading him to settle in first Gloucestershire and later Wales. He became friends with and married into the wealthy Moggridge family, working in partnership with them from then on. He remained close life-long friends with Billingsley, who made him an executor to his will and held the deeds to Perkins' Ashwick land at his death.

more stable, which meant it could be transported more easily than most types of beer. Porter soon dominated the dark beer market. Third, it became illegal for brewers to sell beer direct to the public: there needed to be a retailer such as a publican, which encouraged breweries to buy up public houses, one of the factors which led to the 'tied house' system. All the trends mentioned above, together with increasing numbers and concentration of population, eventually completed industrialization of the industry.[6] And all were relevant in some way to the establishment – and future prosperity – of the Oakhill Brewery.

The process of brewing used large quantities of water: about ten gallons to produce one gallon of beer. A sufficient water supply was essential for the new project. It has been generally thought that the site at Oakhill was chosen for the excellent quality of the water. This is probably only partially correct. Although what water there is in the district is good, the Mendip area as a whole is lacking in surface water.* At least as important was a totally reliable source of abundant water. At the time river water was most preferred for beer – but the only small river nearby, which passed through Shepton Mallet, was already overcrowded and no doubt polluted. Going north from Shepton Mallet and over the summit of the Mendip Hills there is an east-west line of springs part-way down the northern slope towards Oakhill. Being a local man Perkins will have known this, even if Jordan did not.

No information has been found so far as to who the land including the springs belonged to in the 1760s. At that time Billingsley's family held land in the vicinity, very close to but not including that land.† In the early years of the brewery it was probably only necessary for it to obtain water from one or two of the many outlets, but over time quite a number of springs were used, some water piped from as far away as the next parish of Stoke St Michael. Unfortunately, no maps record the existence of the brewery or its main water supply from its beginning in 1767 until as late as 1822 (see the map below), by which time the brewery traded under the name of Jillard & Spencer (the 1822 map was produced shortly after the new partnership's purchase of the site in 1820 – after Billingsley's death).[7] There also needed to be a suitable site

---

*   Rainwater passing down into the limestone dissolves it and soaks through forming cave systems beneath.

†   Land immediately to the south was part of the Forest of Mendip, owned in the 1700s by the Duchy of Cornwall (the crown). This was later Enclosed by the Shepton Mallet Enclosure Act of 1785. The land in question was not included, though at least some parts of it appears to have belonged to the Duchy, which leased it out. One field, possibly the site of the main rising spring, is recorded in 1790 as being held by a freeholder named Wood [Survey of Manor of Shepton Mallet, 1790]

available for the new brewery buildings. In the 1760s Oakhill was merely a hamlet with no more than 50 houses, so not ideal for acquiring employees; it was also rather off the beaten track. However, apart from those needed for distribution, brewing used a relatively small workforce for the size of enterprise. Subsequently, the brewery itself was largely responsible for the growth of the village to over 400 houses within the next century, involving the purchase or construction of a large percentage of them to house their employees. The main brewery buildings must have been purpose built as any existing buildings would probably have been not just inappropriate for brewing but also too small for the project even in its earliest phase.

In a brewery of any size power would be needed to move the water: horse power or water wheel. A fall in ground level would produce a gravity flow making a water wheel possible – just the situation at Oakhill.* The course of the water supply is clearly shown on the 1822 map: pipes took water from the spring in a field uphill, at the southern edge of the map. Though the ponds look separate water could in fact flow between them in a continuous downhill stream, monitored by sluices. The ponds act as both settlement areas and water store. They also expose the water to the air, then believed by some to improve its quality for beer-making.† More than 250 years later, the ponds are still in existence (now reduced to two, with a dam between), as are some of the pipes (there are also large underground water tanks nearer the brewery itself which were built at a later date). Water was led towards pipes (usually then made of wood lined with lead), the line of pipes running underground, then across to near the centre top of the map, they then turn west, across to the brewery buildings. It is believed that even in its early days the brewery used two intakes of water as was then common– one source was for the resulting beer (this water was known in the trade as 'liquor'), the other for the various ancillary processes – though only one source has been found in the fields immediately above the ponds.‡

Drainage from the site would prove to be more of a difficulty than water supply. The site slopes down slightly towards the north, so while the village

---

* Motive power is known to have been provided by a water wheel, though it is unclear whether this was in place from the start or added later. If and when a steam engine was introduced it must have been some time later (not before the last decade of the eighteenth century).

† In 1839 Phelps commented on the Oakhill Brewery water supply: that 'the water after being exposed to the air in a succession of ponds…conduces much to the superior goodness of their porter'.⁸

‡ Brewing tradition meant that any employee who referred to the water due to result in beer as 'water' (rather than 'liquor') was subject to a fine of the same monetary value as for swearing.

was still small waste water would simply have drained away to the north in the usual eighteenth century fashion. Later a better system of drains and culverts took the waste water down into the valley beyond, removing what was called at the time a 'noxious nuisance' (mostly, in fact, due to domestic effluent rather than the brewery itself).

Other prerequisites for a successful brewery included good transport possibilities, as needed for any new business. This would have been another relatively difficult issue: at the time Mendip roads were appalling and no canal was ever built nearby. Horse-drawn waggons were used for supply and distribution, later replaced by drays. Road conditions were a frequent cause of accidents, with the waggons carrying supplies to the brewery or beer away from it (over the years there are several records of brewery employees being killed on the roads). Also important would be easy availability of the other key ingredients. Initially no hops and very little barley were grown locally. Later some barley was grown on Mendip, but the main source of both was from surprisingly far away: barley from Dorchester and hops from Dorset then Belgium. Many breweries of the time – even quite large ones, such as George's in Bristol – did not produce their own malt, buying it in.[10] Oakhill did make its own malt, certainly by the 1790s if not before (thus they needed their own malt-house – and the capital to build it). Later the supply of malt to other breweries became a significant part of their activities. While the heat required for some of the processes would probably have been supplied by wood or 'fern' (bracken), that needed for firing the grain would most likely have come from coal. Ashwick is at the extreme south of the Somerset coalfield, an appropriate type of coal conveniently being then mined within one or two miles of the brewery.

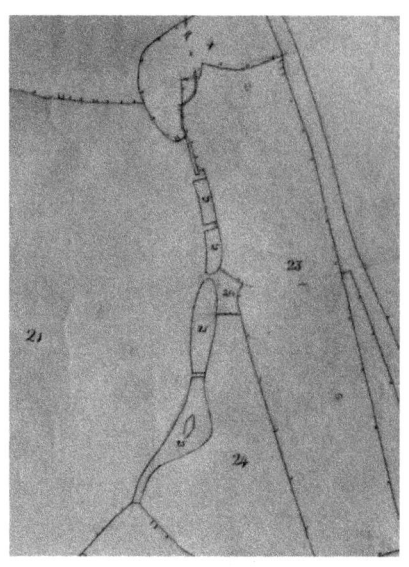

Detail (1) from 'Map of the Estate of Messrs Jillard and Spencer', 1822 showing the ponds[9]

As the founders had no experience in brewing they must have had to find suitable maltsters and brewers who they could trust for the practical part of the process. Though vital for its success we do not know who they were. Neither do we know what variety of beers were produced in the early years.

They will certainly have brewed porter, very much the most popular beer of the day, and one in which the brewery specialised for most of its time. This was a dark beer, almost interchangeable with what later became known as stout. They probably also brewed a pale ale, as they did later, but it is less clear whether they ever produced 'small beer', a weak beer drunk in those days as a safer alternative to water.

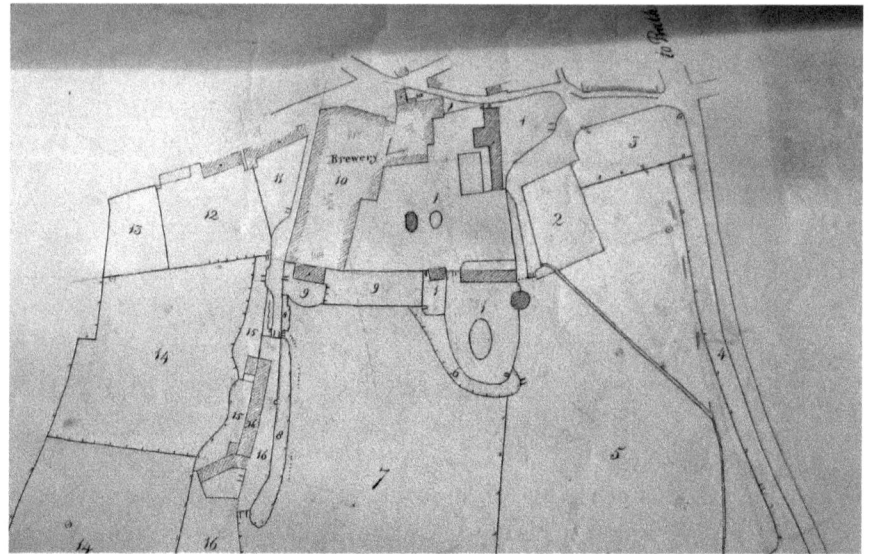

Detail (2) from 'Map of the Estate of Messrs Jillard and Spencer', 1822 showing the brewery buildings[9]

Detail (2) of the 1822 map (above) shows the brewery buildings as they were at that date. To repeat, this is some 50 years after its establishment, but is the first idea we have of the layout (there is no indication that there were any major changes in the intervening years – things were then built to last). The design of the buildings was almost predetermined: brewing requires raising the water and additives to the top level at the start, the various receptacles being positioned at decreasing levels, with the liquor running downwards by gravity as the process proceeds, usually ending in a cellar (though there is no evidence of one at Oakhill). The brewery itself (no 10 on the map) with a forecourt onto Oakhill High Street was one large building with no apparent separation between brewhouse and malthouse, though the L-shape could have been used to accomplish that. The small dark-shaded building attached to the main brewery buildings is The Beeches. This house (which still exists) predates the brewery but was incorporated into it: at different dates it served as either offices or domestic premises for partners or their families. It may have served as an interim base for the founders when they first took over the site, though as it was designed as domestic premises it would never have been useful for the actual brewing processes.

A lane west of the brewery (then called the Drang) would have enabled easy access to the range of stables (no 16) which housed the horses and waggons used for transport. Other buildings in the area were mainly further domestic premises which by this stage belonged to the brewery and were used by partners and employees. This part of the map also clearly shows the pipe coming from the ponds towards the brewery building (from the field marked no 5); the underground route by which it traversed the last section from the corner to the brewery being only recently rediscovered.

Clues showing the brewery's activities in the 1760s and 70s are difficult to find. The Shepton Mallet rate books record James Jordan as in partnership with Richard Perkins at Oakhill in June of 1772; by the next spring they are described as 'brewers'.[11] Later the partners were routinely known as 'common brewers'.* Coincidentally, the year 1772 was marked by a national financial crisis which affected all industries including breweries to some extent. Yet in 1773 the brewery was in a strong enough position to acquire the Oakhill Inn from Mrs Mary James.[12] This acquisition is the first evidence of the brewery's diversification. Within two or three years they also began trading in spirits – and diversified further later.

In lieu of other records, local newspapers dealing with events in Oakhill – chiefly the *Bath Chronicle*, which started in 1760 – have been very useful in discovering what was happening at the brewery. However, not only were breweries secretive about their art (seeing it as a trade secret), they also, for reasons less easy to understand, seem to have never advertised for business in the eighteenth century, even when newspapers became popular and widespread. So, apart from disasters such as fires and accidents, the information that can be gleaned tends to be limited to notices published when there were changes to the business organisation, such as adding or losing a partner.

At some point between 1767 and 1774 George Blake joined the brewery as a third partner, perhaps necessary for financial reasons, whether due to the national economic situation or not (although this particular change is not marked by a surviving announcement). Nothing is known of Blake's previous experience, but things do not seem to have gone well – as we shall see. In 1776 Billingsley himself joined the partnership, with a rather more positive

---

* There was then a distinction between 'brewing victuallers', that is retail brewers of small amounts (for example in public houses), and 'common brewers', the larger wholesale brewers.

outcome. The above description of the brewery's processes and premises gives some idea of the business then and what he chose to invest in. He may well have been involved in its planning and operation prior to that – perhaps simply as a friend and advisor with business acumen, probably intrigued by the substance of the business. His investment displays the entrepreneurial side of his character early in his career.

In March 1776 Richard Perkins left the brewery – not unexpectedly – but so did George Blake at the same time, although at this point we do not know why. The *Bath Chronicle* carried the following announcement:

Oakhill, 8 March: Notice is hereby given that the Copartnership between James Jordan, Richard Perkins and George Blake of the City of Bath, Brewers, is by mutual consent this day dissolved, and that all persons who have any demands thereon, are desired forthwith to send their accounts to Mr James Jordan of this place, who will pay the same, and those who are indebted to them are desired to pay either of the said copartners.
NB the brewery and spirit trades are now carried on as usual by James Jordan and John Billingsley.[13]

Although there are persistent claims even today that his involvement went back further than this, the above notice is the earliest record found of Billingsley's formal interest in the brewery. As noted earlier, the Shepton Mallet weavers' riots did not start until after this, in May of that year. So, either Billingsley had already made the decision to abandon the wool trade before the riot, or he invested in the brewery before deciding to pull out of wool. Perhaps he had been waiting until it was properly established so that he could consider its viability.

These changes must have marked a really difficult situation for the brewery, perhaps even a crisis, related to Blake's departure. Records show that only three months later Blake was being held as an 'Insolvent Debtor', in the King's Bench prison, Southwark.[14] Quite why he was in debt, or how he came to be imprisoned, is not known. Neither is it clear why he was held in London, rather than in Somerset.* Becoming insolvent would have been a shocking situation for Blake – and conditions in the King's Bench were dire – though luckily for him it did not last long: he was released on 29 July 1776, following which he faded into oblivion.[15]

Quite what happened in the 'Blake affair' will probably never be known.

---

* The names of those suing him (Mackrell and Eastman) are not otherwise known in connection with the brewery.

To what extent it impacted on the partnership is unclear, but it must have been a significant — and negative — event. With Perkins also withdrawing at the same time (no doubt taking his investment with him) it could easily have meant financial difficulty for Jordan and for the brewery as a business, although there is no evidence that it did so. As far as is known the cause of Blake's debts had nothing whatsoever to do with the brewery, but the events surrounding it and his tarnished reputation could well have reflected badly on the enterprise.

There is no evidence of money being raised by the brewery by some public method at that time. Other partnerships in the period used shares for investment; for example, Philip George, of George's Brewery of Bristol (then about the same size as the Oakhill) issued shares of £2,000 to eight directors in the late 1780s.[16] But no shares for the Oakhill are known of. Billingsley had already shown himself to be a speculator, so he may have agreed — or offered — to invest before Blake and Perkins actually left, with Jordan realising problems could lie ahead. Billingsley's initial investment in the brewery took the form of land bought 'in trust' together with Jordan (no further funds have been found to have been paid in by him at that time, though he made a greater commitment later). The land was purchased in early 1776 at Ubley, in the Mendip Hills east of Oakhill (a complex process, described in chapter 13). The total cost was £2,371 11s 8d — all paid by Billingsley. The partners immediately used the land as security to raise three separate mortgages totalling £1,200 in 1776, rising to £3,000 in 1777.[17] Most of the land was then rented out, the proceeds presumably benefitting the brewery (with some land, incidentally, being used by Billingsley himself for one of his agricultural experiments).

At the time Billingsley formally joined the brewery he was 29, so still near the beginning of his career. In the years immediately following he became involved in a wide variety of other projects which would have taken up a good deal of his time — agriculture and the Bath Society in particular. He also married. Some of his projects would soon require further significant financial investment: his land acquisition, for example. So we must assume that the brewery progressed relatively healthily following his injection of finance; otherwise, being the shrewd investor he was, he would have taken some decisive action or withdrawn.

It is most probable that Billingsley was never involved in the day-to-day practical running of the company, which was apparently handled effectively by Jordan. While the new partner had extremely varied interests none of his writings show him as paying any special attention to the actual process of brewing or the brewing industry in general, giving it no consideration in his

General View..* Equally, though, he is unlikely to have been the proverbial 'sleeping partner'. His tendency was to take a very active interest in all his projects, particularly in financial matters, as evidence shows was later the case here.

It might be hoped that some idea of the state of the business and its assets at the time could be gained from tax records. The first available records are from 1782, when 'Jordan & Billingsley, Brewers' paid local rates on two public houses in the village of Oakhill (the White Horse and the Moon). But, oddly, no rates seem to have been paid on either the Oakhill Inn or the brewery itself. In the 1798 land tax records for Shepton Mallet tything, Jordan and Billingsley paid tax for 'stock in trade' – rather more than most businesses in Shepton paid at the time, though theirs is not the highest charge.[18]

While tax records do not reveal a great deal, it is known that this was a difficult period for breweries and for businesses in general: 'with the year 1788 commenced a series of bad harvests and a long period of distress'.[19] Political and economic issues, and war, contributed to the difficulties. Many breweries were struggling, due to fierce competition as well as taxation and market conditions. Despite their efforts at diversification George's of Bristol, for example, experienced real financial problems and did not turn a profit for several years. By moving into trading in wine and spirits, plus beginning to buy up public houses (soon owning a large number), the Oakhill Brewery had diversified early and seems to have fared rather better.†

Collinson's well-known *History and Antiquities of Somersetshire*, written in 1791, reports that Oakhill was 'now only famous for a large Brewery carried on with great reputation by Messrs Jordan and Billingsley and both these men have good houses there'. Billingsley's house was of course Ashwick Grove, but where Jordan lived has always been something of a mystery. Some have thought he lived at The Beeches – very much on top of the job – others that he lived in a house in another part of the village. However, detail (2) of the 1822 map (above) shows the most likely contender. This is the large dark-coloured building, facing east, in the top centre of the map and to the east of the brewery. It has now been established that this house was called Oakhill Cottage (photograph below). Almost certainly Jordan lived there, and very likely built it. At some point the extensive grounds of what was Oakhill Cottage were set out as 'gentrified' landscape, as was then fashionable. One intriguing part of

---

* On the other hand, he seems to have regarded 'cyder making' as sufficiently important in Somerset to describe the process.

† Billingsley also purchased the Bell Inn at Fromefield, Frome, as an individual in 1792 – which no doubt sold Oakhill beer.

*Oakhill Cottage, predecessor to today's Pondsmead
James Jordan's home at the time of Jordan and Billingsley*[20]

this is the still-surviving grotto complex (now grade II listed), thought to be relatively unique.*

We know very little about the management or operation of the brewery during the years after Billingsley joined until the turn of the century, apart from the following bit of information. By 1790 one their most valued employees had begun work there. Mr Nicholas Reynolds was 'the Clerk, in which situation he served for upwards of 20 years…with the strictest honour and integrity'.†[21] This enabled some continuity on the financial side. Evidently he was much valued, the senior clerk in a brewery usually managing its day-to-day finances and also being effectively the general manager. Reynolds seems to have been considered a trusted friend as Billingsley remembered him in his will (although the former died in November 1810, aged 63, shortly before Billingsley himself).

The first decade of the 1800s was another time of great change for the management. James Jordan had retired by 1803. He was only about 58, so by our standards quite young to retire. He passed at least some of his share

---

* The extensive grottoes were built into the dam between the two ponds. They are believed to have been commissioned by one of the co-owners of the brewery, either Jordan or his successors at Oakhill Cottage, the Jillard family, at a date yet to be determined. Billingsley himself is unlikely to have had a hand in it.

† Reynolds is described elsewhere as being 'an officer of the Excise' (in the presbyterian register of baptisms, so evidently also another presbyterian).

in the brewery to Billingsley (see chapter 9). He moved away from Oakhill, leaving the house available for the next partner, then lived to the ripe old age of 85 (dying in 1830). Meanwhile, Billingsley's main interest had been turning inexorably towards agriculture, so that by the 1790s he was regarded as an expert on the subject, although he continued to be a partner in the brewery. In 1803 he moved to Bath, probably for health reasons. But while Jordan had stopped being responsible for the brewery's rates by 1800, Billingsley was still involved in their payment for several years past this date. Around the time Jordan retired William Peard Jillard (1768-1849) joined as a new co-partner with Billingsley, the business becoming known as 'Billingsley & Jillard', the former now the senior partner. It is most likely that Jillard took over the practical day to day running of the brewery from Jordan.*

Intriguingly, it was during this time that 'Billingsly [sic] and Jillard' are recorded as operating as 'Bankers' in addition to brewing, another form of diversification.[22] The connection between brewing and banking is said to have been relatively common at the time, with brewers moving into banking and bankers moving into brewing. It has been remarked that 'the distinguishing mark of the [brewing] entrepreneur was non-brewing activity'.† The link was due to breweries being one of the businesses which had larger amounts of capital invested than many others, leading to greater stability.‡ At the Oakhill Brewery it is most likely Billingsley will have led this activity (even though he was in Bath at the time), but whether it had started during Jordan's tenure or only once Jillard came into the business is not known. Unfortunately, once again, no records can be found detailing their activities in this sphere. No doubt Reynolds was responsible for the day-to-day part of it. In the days when banks were precarious and subject to frequent failure breweries often operated

---

* William Peard Jillard was the son of the Rev Peard Jillard (1731-1799), a nonconformist minister who served at the presbyterian chapel in Shepton Mallet from 1754-70. He would have already been well-known to all the people mentioned here. Jillard junior was educated for the ministry and may have briefly served in it. He had several children; two sons died in infancy and another two tragically in their twenties, one of whom had worked briefly at the brewery. Fortunately, another son remained who succeeded his father in the partnership.
† [Mathias, 1959], As an example, Sir Benjamin Hobhouse, a Bath banker and president of the Bath and West Society (who gave the eulogy after Billingsley's death), was also a partner in Whitbreads. Later members of the Hobhouse family became partners in the Oakhill Brewery.
‡ The cloth trade was another candidate for bank activity: it is recorded that in the early 1790s three separate banks had recently been established at nearby Frome to help service the trade [Universal British Directory, ca 1794].

as recipients of funds for family and friends, as well as for early versions of what are now thought of as 'Friendly Societies'– in fact they lent to anyone wanting fast access or a good return for their available money and not wishing to put it into government bonds. Brewing bankers lent most often to those in associated trades (such as maltsters, publicans and so on) either as mortgages or to help them in their business. This kind of activity was particularly useful in rural areas like Oakhill, with little access to other banking options. Even if it was on a relatively minor scale the brewery must have been seen as sufficiently successful and secure to have been able to offer this service.*

Two other partners also joined in the early 1800s. The first of the Spencer family of Shepton Mallet began their long association with the brewery. They too were nonconformists – seemingly a prerequisite for partnership in the business.† John Spencer (1765-1851) had been a Shepton publican, running the Black Swan public house in the market place before coming to Oakhill. Already recorded as a 'Brewer', he was probably the first investor to bring personal expertise in brewing to the business.‡ Presumably the partners had been employing one or more skilled brewers for the actual process; now they were no longer totally reliant on employees for that expertise. Beyond this, and apart from the fact that Billingsley was almost certainly the lead in financial matters, the various roles of the three partners within the business are not clear. Edward White had also joined the partnership before September 1810. Little is known about White, except that he was a partner until 1820.§

The date of Spencer's entry to the partnership is not marked by surviving documents, circumstantial evidence only gives it as sometime between 1803-9, when Billingsley seems to have begun winding down his own involvement. Spencer's son John Plummer Spencer, and *his* two sons, John Maitland and Frederick were all partners in their turn, managing and growing the business very effectively for the rest of the nineteenth century. By 1809 the brewery had expanded sufficiently to have an 'Oakhill Beer and Porter Warehouse' in Back St, Bristol.[24] This, they announced, served customers with 'genuine strong Beer and Porter'. By this time the brewery also owned public houses in

---

* In 1918 the Shepton Mallet Journal reported that paper money guaranteed by the brewery was still in existence.[23]

† There was a very strong connection between brewing and nonconformism (many of the largest London breweries were owned by quakers, for example).

‡ It is possible that George Blake had practical brewing expertise.

§ Jillard described White as his 'Friend' in his Will. Jillard's son Robert Haskoll Jillard replaced White as the third partner in 1820 (Robert died of a 'brain fever' in 1826 and was himself replaced by his brother William Vernon Jillard).

Bath and elsewhere in Somerset. England had then been at war with France for some years, making national conditions difficult in many ways – but it did not noticeably lessen demand for beer.

Billingsley moved back from Bath to Ashwick in 1808. He withdrew from the brewery business by September 1810 (no reason is given – possibly other commitments, or just simplifying his financial affairs). This produced the following announcement from the *Bath Chronicle*:

> Oakhill, 29th Sept, 1810
>
> Notice is hereby given, that the Copartnership between JOHN BILLINGSLEY, WILLIAM PEARD JILLARD, JOHN SPENCER, AND EDWARD WHITE, under the firm of 'JILLARD SPENCER AND CO', is this day by mutual consent Dissolved. – All Debts owing by and to the Partnership will be paid and received by the said William Peard Jillard, John Spencer and Edward White at the Brewery at Oakhill, aforesaid.
>
> Messrs JILLARD, SPENCER & WHITE, Proprietors of the OAKHILL BREWERY, respectfully inform their Friends and the Public, that they carry on the Business at Oakhill as usual [25]

At some point prior to this date they had already changed their name to 'Jillard, Spencer & Co', omitting the Billingsley. Billingsley died just over a year later. William Peard Jillard and John Spencer, then the two main partners, continued to grow the business. They now described themselves as 'Brewers and Spirit Merchants'. By this stage the enterprise had become a substantial operation.

The hey-day of the brewery was the years between about 1820 and the first world war. Before the outbreak of war it was a very successful concern: winning prizes; owning a large number of public houses dispersed through the region; running its own railway (probably one of only two brewery railways in the country); having a nation-wide distribution system; and reputed to have sold more stout than Guinness! The brewery had an enormous impact on the village, with succeeding partners all keeping to the philanthropic tradition established by the early nonconformist co-owners, with great benefit to the village and surrounding area. The brewery was forced to close in 1925 following a major fire, after which both the business and site were taken over repeatedly and used for a number of brewing related concerns for many years thereafter. What remains of the buildings have now been converted into domestic premises.

Within the folk-lore of the area and that of beer enthusiasts, the name of John Billingsley has been given a very central role in the Oakhill Brewery: it is claimed that he was the founder and sole-owner. Neither was the case. Nor is it likely that he was as crucial in the day-to-day running of the business as some have thought – apart, perhaps, from the monetary side. Billingsley was not a brewer, but he would have been involved in the financial running of the business, encouraging its diversification and probably central to its banking operation.* He was an entrepreneur who could see promise from the time of the brewery being a relatively new concern, though for most of the period he was a co-owner he was extremely busy with other projects. He maintained his investment until very shortly before his death, helping to lend financial stability to what became an extremely successful enterprise.

---

* He may also have been involved in the various water engineering projects, which at some point included the formation of cascades over the dam (their construction so far not dated).

6

Agriculture and the Bath and West – Part 1

Agricultural improvement was arguably the main focus of Billingsley's working life. According to his own account he began being active in farming in the early 1770s, while still in his twenties and working as a clothier. He seems to have already felt that the methods then used were not as effective as they should have been, as he immediately began experimenting with ways to improve the land and its produce.[1] Before describing his activities in this it may be helpful to look very briefly at the strands of change in eighteenth century agriculture and at the origins of the Bath Society which also became central to his life.*

It was a time of great change in agriculture. A complex picture, the transformation was due partly to the introduction of new crops; to the widespread Enclosure movement; was affected by economic and political factors; and took place over a relatively long period of time. The changes were also influenced and encouraged by a large number of individuals and groups – John Billingsley and the Bath Society among them. Agricultural societies began to be set up in England from the 1770s.† These societies soon became one of the chief means through which the new ideas spread: through discussion, demonstration and encouragement, for example by offering 'Premiums' (prizes) to those who employed the most novel or outstanding methods. The best way of ensuring change was soon seen to be by spreading the word to the largest number of practical farmers.‡ Agricultural improvement – and

---

* The Bath Society later became the Bath and West Society and is now the Royal Bath and West Society.
† Following the establishment of the Royal Society a century previously, scientific and literary institutions and societies had become popular: agricultural societies were a part of this wider movement.
‡ Before the advent of agricultural societies, Thomas Coke claimed that among farmers new ideas travelled at the rate of about one mile per year!

agricultural societies – became fashionable.

The prime mover for such a society at Bath was Edmund Rack (1735-1787). Rack was an intriguing character: a quaker, born in Norfolk, by trade a draper who worked in Essex before relocating to Bath. There he quickly settled into the city's literary and social scene and cultivated good connections.* He had always had a keen interest in agriculture. Having moved to Somerset he found that, in his opinion, agriculture in the west country was much inferior to that in Norfolk, then an acknowledged leader in farming methods. Believing that the west country needed to catch up, by 1777 he had published a series of letters on agriculture. The response to these encouraged him to set up a society. The Bath Society was established in 1777, one of the earliest agricultural societies in the country. Following advertisements in the *Bath Chronicle*, Rack assembled a group of 22 interested men, who met at York House in the city of Bath on 8 September, 1777.² The group included, apart from Rack himself, many prominent Bath citizens and local landowners – but none with any personal agricultural experience.† 63 Subscribers, including the Earls of Ilchester and Salisbury and Dr Joseph Priestly FRS, put their names forward. Having decided to proceed, the first resolution was to elect a secretary. Rack was chosen unanimously. The group assembled itself into a committee that met once more to prepare 'One General Plan to lay before the Society at its First Meeting'. The society was to cover the cities of Bath and Bristol and the counties of Somerset, Wiltshire, Gloucestershire and Dorset – an ambitious target, but one which was more than reached almost immediately. Soon, through its subscribers and corresponding members, it was actively linking interested men country-wide.

Despite the frequent assertion that Billingsley was a 'Founder Member' of the society, although he had been invited to attend he was not present at either of the first two sessions mentioned above, nor was he a subscriber. Billingsley was only 30 at the time, he had just joined the Oakhill Brewery and may well have had competing commitments on the days of the meetings. But he must have paid attention to what had transpired as he did attend the third meeting on 13 November, 1777. This was the 'First General Meeting':

---

* Rack probably chose Bath for its intellectual scene: having widespread interests in both poetry and science he produced copious publications on a variety of different topics. He was co-founder of the Bath Philosophical Society in 1792 and also took on a gruelling project for the Rev John Collinson, who then published *A History and Antiquities of Somerset* (1791), using Rack's information but without crediting him.

† Initially the society had rather broad aims, including manufacturing and chemistry for example, but settled over time to focus on agriculture.

his having attended then will presumably be why he is often seen as a founder member. His fellow members certainly saw him as one.

Throughout his close lifelong association with the society Billingsley was never a regular attendee at general meetings (held bi-monthly) but was always very actively involved in society matters. He usually attended the important annual meeting in December but at other times he was mainly only present at relevant committee meetings or when business involved him personally. This pattern is understandable considering his many other pressing interests, coupled with the state of the roads and the fact that his home was several miles from Bath: attendance would have meant a journey of four or more hours each way, often involving an overnight stay. And as Billingsley had already begun to be active in practical agricultural improvement before its establishment, the society itself cannot have been the prime motivator for his efforts in this direction. Even so its activities must have encouraged him from the start, as shown by the many letters and reports he wrote for it on his various projects. It is the society's *Letters and Papers*, plus minutes of its meetings, that provide us with most of the early information as to his activities in agriculture as a whole. Without the Bath Society publications we would know that he farmed large tracts of land, but would not know exactly what he did on any of it until much later, when he published the *General View..* (1794), which even then was more concerned with county agriculture than his own farming methods.

At the first general meeting there were 36 men present, including Billingsley (there appear to have been no female members at all during his lifetime). The increase in numbers, together with a large amount of correspondence (some from distant places) demonstrates the interest the society was already gaining. Lord Ilchester was designated president and twelve members elected as vice-presidents, Billingsley not among them. He may well have been seen as a junior member at this point, although by the start of the society he was already preparing his land and recording his results in a much more thorough way than most of his contemporaries, then building on this experience to inform his views – which were always strongly and confidently held.

The aims of the society included the intention to 'excite a spirit of enquiry which may lead to improvements not yet known; and to bring speculation and theory to the test of accurate experiment'. The secretary reported that they had 'a desire to bring to the western counties the obvious advantages which those in other parts of the kingdom enjoy where societies of a similar kind have been formed' – presumably with East Anglia in mind.

In the early days the society met at Rack's house, in rooms reserved for their meetings. Rack was clearly a very efficient and competent secretary:

despite his many other activities he was extremely well organised and hard working. The sheer quantity of correspondence is striking, some of it needing a knowledgeable and/or diplomatic response – some members could be very awkward and time consuming. As remarked by Hudson in his history of the society 'the Secretary's job was no sinecure'.[3] Rack's capability was acknowledged by the members: each year thereafter they voted unanimously for him to continue as secretary and thanked him for his efforts.

A number of committees were set up at the first general meeting, including one for manufacturing and commerce, for which Billingsley was the first person nominated (no doubt due to his reputation in the wool trade); and one for agriculture and planting, for which his name was not put forward (perhaps his interest in agriculture was not yet sufficiently recognized). Each committee issued its own list of premiums annually, in a range of appropriate areas. Premiums were also given to 'agricultural servants' nominated for loyal long service, and those who successfully raised large numbers of children without the financial help of the parish – the so-called 'Servitude Premiums'.*
Lists would be reviewed and altered annually, the subjects chosen reflecting those most occupying the minds of the society at any one time.

Billingsley attended the society's first annual meeting, in December, 1777, when his name was added to the committee for agriculture and planting. He became increasingly involved in this committee, but was never especially active in that for manufacturing and commerce (by this stage he had withdrawn from the wool trade). All decisions made by the various committees were presented to the annual meeting, discussed where necessary and ratified. Over the years the annual meeting came to involve sessions on two or three days, with the main committees meeting in advance, often running out of time and adjourning for a second session, or even a third. From this time on Billingsley became a close collaborator with Rack and almost certainly a friend, the two men exchanging numerous letters between meetings and later travelling together.

The society moved fast. By December 1777 it had 300 full members; its first book of premiums was already 'in the Press'; £240 had been lodged in the bank by February 1778; the next month the society's lending library was launched. It was a very successful start. Letters were read from corresponding members; models and drawings of machinery were viewed; samples of tartarian oats and grass seed examined; cultivation of turnips and rhubarb

---

* Although there were several servitude premiums available every year no servant of Billingsley's was ever awarded one – the only man he put forward was disqualified as having filled in the form incorrectly.

discussed at some length – particularly rhubarb, which seemed to obsess the society for some years.* Unlike the majority of members Billingsley showed no interest in rhubarb. By June 1778 the shape of the society's business had settled down sufficiently for it to begin to publish notices on the correspondence and proceedings of the society. These could be seen monthly in the *Farmer's Magazine*. Billingsley is not known to have contributed.

Minutes of the general meeting in October 1779 record that 'Mr Billingsley gave notice that he would claim the Premium for raising Carrots', his first foray into competition in agriculture. Assessors were asked to view the crop and report back at the next meeting. Only two days later it was agreed that he would receive his first premium of 10 guineas, to be awarded at the annual meeting in December. Billingsley's choice of carrots is curious, as today the land then used by him is regarded by local farmers as unsuitable for that crop; none are grown commercially in the area. But evidently he was successful. He had wished to also claim a premium for growing cabbages, though as in the opinion of the assessors the result was 'so mixt [sic]' he was asked to withdraw the claim. Both these crops, incidentally, were grown for animal feed rather than human consumption.

Billingsley was added to the list of vice-presidents at the annual meeting of 1779. From that time on he frequently chaired committees and occasionally even the annual meeting itself. The society's main income was from subscriptions but there were always many outstanding; in some years this caused the society financial embarrassment in view of the commitment to premiums already offered. In common with others from outlying districts Billingsley was tasked with collecting subscriptions, in his case from people in the Shepton Mallet area. He also became a member of the committee to revise rules and orders. It is not clear whether members of the society volunteered, or were volunteered, for the numerous committees, but Billingsley was very often on the list. He frequently served on temporary sub-committees (always simply described as 'Committees') which were often set up for special purposes such as to assess claims for premiums. For such a busy man he was either a very willing or a very popular choice – a trend which continued. However, the committees may not all have been onerous: according to the records a temporary committee might be set up during a meeting, then its members would simply withdraw to deliberate and report back on the same occasion, or at the next meeting after a brief adjournment, or following an inspection (as happened here with Billingsley's carrots).

---

* Rhubarb was then used for medicinal rather than culinary purposes, constipation among other things – another obsession of the time.

At the annual meeting of 1781 Billingsley was awarded a further premium 'for 15 acres of Carrots with near 40 tons of produce per acre – 10 guineas'. In the meantime his previous success for eight acres of carrots had already led to a paper appearing in 1780 in the first volume of the society's occasional publication: *Letters and Papers of the Bath and West Society*.[4] This volume also contained a number of letters from Norfolk and a considerable amount on rhubarb, in addition to Billingsley's 'Account of the Culture of Carrots and thoughts on Burnbaiting on the Mendip-hills'. The committee setting up the premium had set the rules for the trial for carrots, such as that the crop should be kept clean by hoeing, but had not specified details of the preparation of land – thoroughness of the latter being one of Billingsley's fortes – nor had it specified the records to be kept. The article shows that Billingsley had already begun preparation of that particular land in 1776, before the establishment of the society. When representatives of the society visited to inspect the crop they requested that he should give details of the culture, expenses and produce which he had recorded, resulting in the paper.

In 1776 he had limed the field before planting turnips; in 1777 he grew barley; in 1778 after liberal use of horse-dung he grew scotch cabbage (which he 'recommend[ed] in the warmest manner').* In 1779, after very thorough preparation of two ploughings each followed by harrowing, he sowed eight acres of red sandwich carrot seed in drills, kept clean by hoeing.† The carrots were used to fatten pigs. The costs of all these activities (going back to 1776) are recorded in the paper, with a profit of £47 10s from the eight acres. The paper demonstrates that by the latter part of the 1770s Billingsley's characteristic way of working and record keeping was already well established. Burnbaiting, at first sight an unrelated topic, was discussed in the second half of the paper: this practice of paring up turf and burning it in order to produce ash to fertilise the soil, was a method then frequently used in the south-west. Billingsley described having tried it, but, according to a letter later in the same volume from 'a gentleman from Norwich', the trial resulted in 'Burnbaiting condemned'.

Another article by Billingsley was soon published, in volume II of the *Letter and Papers*, 1783.[5] This again concerns carrots: 'On the Profit of Carrots and Cabbages'. Here he sets out his expenses and return, saying he tries to

---

* Billingsley says in the paper: 'the field in which the carrots were raised was a few years ago part of the Forest of Mendip… the soil a gravelly loam of a good depth'. It can be identified as a section of recently enclosed land at Ubley, bought in 1776 as part of his investment in the brewery.

† He was a strong advocate of hoeing. He makes no mention in the paper of pests or diseases, although carrot fly, one of the main pests of carrots, is endemic and a serious risk today.

avoid exaggerating the value of crops, taking an average calculation, not best or worst. His effort to avoid exaggerating the amount of return is claimed frequently in his writing, and from later evidence it does seem to have been his genuine practice. In the same volume is a note from Arthur Young, the agriculturalist, to the effect that 'amongst those whose results are published Mr Billingsley ranks the foremost'.* However, Young goes on to say that when he had viewed the farm more recently Billingsley 'had given the culture up, there was not a carrot upon it, from which we must conclude there was a deficiency in value'. Perhaps, like farmers in our own day, he had concluded that the area was not best suited to carrots.

By 1784, if he attended general meetings Billingsley was usually taking the chair (he seems to have been regarded as already skilled in this task), and was also acting as a judge for claims of premiums. He was himself awarded a premium of 10 guineas for growing potatoes – 70 acres, from which the produce of six acres was chosen as being representative.† This was reported in the newspapers, which gave some detail as to his method: two ploughings from an oat stubble; harrowing and liberal dunging; planting of the seed potatoes in beds five feet wide; with alleys of two and a half feet between; the seed then covered with earth. The produce amounted to 350 bushels per acre (a bushel weighing 60 lbs). The previous year he had achieved an even better result of 389 bushels per acre. He published an account of their culture in volume III, the title in full being: 'Culture, Expenses, and Produce of Six Acres of Potatoes, being a fair part of near Seventy Acres, raised by John Billingsley Esq', for which the Society's premium was granted him in the year 1784.[6] He also included details of the profit from this venture.

In 1783, having paid 'a very high price' for Dumfries seed potatoes his profit was double that of the black oats he had grown on the same land the previous year. In 1784, growing 70 acres of potatoes and using seed potatoes 'of the white sort', he achieved an even greater profit for a representative six acres (£73 11s). It is interesting to note that in 1783 Billingsley had used the most expensive seed potatoes from Scotland – presumably regarded as the

---

* Arthur Young (1741-1820); best known as an agricultural improver, was also a social and political observer, who travelled widely and wrote prolifically. Although he lived at a distance he contributed regularly to discussions by letter and in print as a corresponding member of the Bath Society. He was a critic of farming practice (including evaluation of Billingsley's work), although an on-going criticism of him was that he had not been successful in his own farming efforts and was more a publicist than practical agriculturalist. Young was also influential later as secretary of the Board of Agriculture.

† Evidently the 10 guineas was converted to a pair of silver cups [information from Mrs W Moger of Bath, a descendent, reported by Robin Atthill 1955].

best. Something must already have been known about potato disease being less prevalent there. Although they knew little about the cause of most diseases in the eighteenth century we now know that altitude inhibits the growth of several potato diseases. At the time the most concerning was 'the curl' (much discussed within the society), a group of infections now understood to be caused by viruses. Potato blight, a scourge of the crop today caused by a fungus, had not yet reached these shores. Billingsley also published his results on fifteen acres of Scotch cabbage, which had recently gained him a further premium of 10 guineas.

In addition, volume III published a debate between Billingsley and Thomas Davis Esq of Longleat, concerning the relative merits of dairy and arable.* Although by then he did own a dairy herd Billingsley was much less active in dairy husbandry than in arable and pasture, and much less experienced in dairy than Davis. Nevertheless, he entered the discussion in his usual assured manner. He considers it impossible to have a premium for dairy management as there are so many different considerations outside the control of the farmer. Cheese was already the dominant dairy product for the west country. As to whether milk might be skimmed in order to make more butter but still make profitable cheese, as Davis suggests, he is unconvinced. Davis claims that cheese made with *no* cream (as was usual in Dorset) loses only 5s or 6s in value per hundredweight, whereas Billingsley disagrees and puts the loss at 8s. With respect to profitability of arable versus dairy, Billingsley claims it can easily be seen that the latter is superior, as dairy farmers are 'much more likely to pay their rent on time'. The most notable comment in the exchange, though, comes at the end: the first evidence of Billingsley's important conclusion that the ideal proportion of pasture to arable in most of the south-west should be three of pasture to one of arable, a view which he held in principle and practised thereafter.†

In April 1785 the society arranged the first of its many local ploughing matches. One newspaper described it as follows:

---

* Thomas Davis (1749 -1807), steward to Lord Bath at Longleat between at least 1784-1798. Important locally, a senior member of the Bath Society and contemporary of Billingsley, he developed an experimental approach to agriculture, many of his papers being published by the society. Davis was successful in improving the Longleat estate's grassland and its stock, especially sheep and dairy animals. Commissioned by the Board of Agriculture to write the *General View*.. for Wiltshire.

† Where arable was used (as opposed to permanent grassland), a four-year rotation with three years of grass and one of arable was still the norm in the area until about the year 2000.

> Last Thursday Mr Henry Vagg of Norton Down, with the Norfolk plough and 2 horses without a driver, ploughed 2 acres statute measure in 5 hours and 38 minutes – and Mr Billingsley the same quantity of land with his double-furrowed plough and three horses, in 3 hours 50 minutes.

Another ploughing match held the next Monday, between Mr Thomas Robins of Bowl-down near Tetbury and Billingsley was won by the latter:

> Mr Billingsley's double-furrow plough did an acre and a half in 3 hours, the greater part of the time with 3 horses. Mr Robins used his own plough with 2 horses without a driver and ploughed something more than ¾ of an acre in the same time then declined any further contest… [following this, Mr Billingsley's ploughman] ploughed several furrows without being held at all by the ploughman, who only drove the horses, to the astonishment of several farmers, who declared the work to be most competently performed and that without having seen it they could not have believed it possible to be done.[7]

Impressive. But this was not the end of the story. Further details emerged through a series of letters to the newspapers, their authors hiding behind aliases as was common at the time. The first letter came from 'A Lover of Agriculture'. This claimed the horses used by Mr Billingsley were of very high value, while those used by Mr Vagg were obviously much inferior and the plough not in good repair, plus a third plough (not mentioned in the previous reports) was used with only one horse. Mr Billingsley's plough had not been taken up by farmers though known about for more than twelve years, while that of Mr Robins was only six months old but was now being rapidly adopted. Other information shows that the third plough belonged to Mr Tugwell, another leading member of the society (although he is not named in the letter). An elaborate 'Digest' (calculation of distances) was created to compare the amount ploughed, to the detriment of Mr Billingsley. This letter was rapidly followed by one from 'A Member of the Bath Society'. He objects to 'A Lover of Agriculture' on several grounds – the digest is rejected; the Norfolk plough is known to be superior on heavy ground and the double-furrow on light, Mr Billingsley being thus disadvantaged by the land being 'heavy rather than light'. Mr Billingsley had offered an exchange of horses, which was rejected, the bet they had started with was set aside by his opponents, and so on. More important, the letter makes clear that Mr Billingsley had never claimed the invention of the double-furrow plough (as insinuated) and always credited its maker, Mr Brown of Tamworth, blacksmith. The controversy died down, but evidently simmered. Interest in ploughing matches increased. Billingsley's proudest ploughing moment was yet to come.

The double-furrow plough is one of the things for which Billingsley is best remembered: it was his most favoured implement, particularly on light soils. In later years he entered many more ploughing matches using this type of plough and nearly always won. In some circles Billingsley's name became almost synonymous with ploughing: his portrait even shows the double-furrow plough in the background. Soon after the 1785 match he deposited a model of the plough with the society for members to view. The above report shows that he had used horses on that occasion, but he was a strong believer in oxen as superior for the purpose, also seeing them as being more cost-effective in terms of feed and value of carcass after their working days were over.

The prime importance of ploughing in agriculture was well recognised at the time: in the introduction to volume II of the Society's proceedings (1786), Rack had written 'it being universally acknowledged that in the whole circle of agricultural practice there is nothing so interesting to the Farmer than to plow [sic] cheap and well'. In future, in order to stimulate improvement the Bath Society would arrange frequent matches, says Rack. Billingsley is reported (in volume III) as having 'justly observed' that the counties covered by the society are so far behind in ploughing. 'Vigorous attempts', Rack claims, have been made by the society in encouraging farmers to use the Norfolk or other ploughs rather than their current ones. Meanwhile, Billingsley kept a dignified silence with respect to the comments made by others on his performance in the ploughing matches. In volume III, though, he says that while many farmers receive daily ocular demonstration of the inferiority of their own ill-constructed plough, the fault lies more with the ploughman rather than the master, whose indolence induces him rather to accommodate the plough to the man, than to accommodate himself to the plough.[8] A note from the editor follows, to the effect that the society thinks him correct, and will introduce a premium to induce ploughmen to adopt and properly use the Norfolk plough. It must have been disappointing to Billingsley that his favoured double-furrow plough was not the one chosen.

In 1786 Billingsley made an important – and relatively rare – journey. He travelled with Edmund Rack on an agricultural tour of Norfolk, Suffolk and part of Essex, being away for some of the time between the middle of June and end of September of that year. It is odd that no report can be found as to the destinations or outcomes of this tour. This is particularly unfortunate as it must potentially have been very useful for both men – and for the society as a whole. Rack will have had contacts he could use to survey the various farms which were in the forefront of innovation and excellence. It is not known quite how long the two were away, nor how arduous the journey was. But as an asthmatic it could well have compromised Rack's health. Asthma can be unexpected in its effects, especially as there was no effective medication for it in those days. (Billingsley was also, by co-incidence, an asthmatic, but there is no report of his suffering its effects at this stage of his life.) Perhaps ill-health was the reason Rack did not write up the results as would have been his usual practice. He was still busy working on Collinson's book at the time, and quite probably exhausted by his efforts. In the event he returned from the tour of East Anglia only a short time before he died.

Billingsley attended the annual meeting and dinner in December 1786, at Bath's White Hart Inn. There was plenty of time for networking, something he was skilled in. The secretary, however, is unlikely to have been present; he was already ailing by then and soon seriously ill. By the new year Rack was aware of his approaching death: on 10 January 1787 he wrote to Collinson, 'for 10 days I have been so weak I could scarcely hold my pen. I must resign myself to the fury of the storm which will soon hide me for ever'.[9]

Rack died on 22 February 1787. It was an incalculable loss to the society. It must also have been a personal loss to Billingsley who had worked so closely with him. There was surprisingly little mention of Rack's demise within the society and no obituary in the Bath Chronicle, simply a notice as to his death. Within a month his effects were put up for auction. Plans must have already been made for a replacement secretary: at the same meeting that announced Rack's death it was also agreed that William Matthews would take over his post. The imminent ploughing match was cancelled. Billingsley did not reappear at general meetings for most of the rest of that year. It was not until the annual meeting in December that he attended again.

7
Agriculture and the Bath and West – Part 2

Following Rack's death in February 1787 William Matthews, already a member of the society, was elected secretary, serving in that role for the next twelve years. Like Rack, Matthews was a quaker. He ran a business selling seed, agricultural implements and machinery and had been an enthusiastic member of the society since its inauguration. The society then moved its headquarters to Hetling House in central Bath and he ran his business from there, buying the house some time in the 1790s.*

Matthews settled into the task, but it was more than a year before he was able to rearrange the cancelled ploughing match.[1] This important event was finally held on 27 March 1788 at Barracks Farm near Bath. It is believed to have been the first national ploughing contest ever to take place. There were six competitors.† Two withdrew as their ploughs were not up to the task, the plough of a third (Lord Weymouth) suffered damage against an unexpected rock, leaving just three to compete. First prize went to John Billingsley, using a double-furrow plough with six oxen; in second place Farmer Sully of Midford, using a single plough drawn by six horses; and third, Mr Thomas of Keynsham, with a light swing-plough drawn by four small oxen (none of them used the then most popular Norfolk plough). There appears to have been no altercation following this event, either over type of plough or beast to draw it: Billingsley was vindicated in both his choice of plough and of oxen over horses.

His reputation climbed. Atthill claims that ploughing was his 'greatest single contribution to the improvement of Somerset agriculture'; also that 'the occasion of his greatest personal triumph was the introduction of the double-furrow plough' (though this judgement is debatable – see later).[2]

---

* This grade I listed house (now known as Abbey Church House) is one of the major buildings of Bath; reputed to have been lived in by Alexander Pope for a short period.

† While it was offered as 'national' in fact five of the entrants were from Somerset and only one, Lord Weymouth, from neighbouring Longleat in Wiltshire.

The importance of ploughing remained near the top of the society's agenda for some time, being a constant source of discussion and experiment, the question as to whether horses or oxen were preferable as draft animals less so. The society began a long series of annual ploughing competitions. Having produced the usual report Billingsley was then asked to rewrite it, for a reason not stated. No revision seems to have survived. Shortly after this he engaged in an open discussion with Mr Cooke on the relative merits of deep and shallow ploughing – Billingsley being an advocate of the latter – so another trial was suggested for March 1790, 'to be held at Mr Billingsley's'. Unfortunately, there is no record of the result of this either. Early evidence that Matthews, though relatively well-organized, was not quite as careful as Rack.

The new secretary wrote to Billingsley very frequently – even more frequently than had Rack – sometimes about seed, but more often on society business, especially to request his opinion on decisions to be made. For example, he wrote on three occasions about the annual meeting for 1790. It seems that by this time Billingsley was viewed as one of the most knowledgeable and useful members of the group, with Matthews regularly using him as a sounding board.

Shortly before the 1790 annual meeting Billingsley's crop of corn was examined with a view to his claiming a premium. By September 1791 his report on the crop was due but had not yet been received. For a man usually so efficient this tardiness must have been occasioned either by illness (of which no mention at the time) or by pressure of events (it was an even more than usually busy time for him). He had complained to the society about his grassland receiving improper comparison with Mr Crook's, his competitor for the award: he felt his land was much less favourable than that of Crook, which needed to be taken into account. This was evidently agreed to by the committee. At the annual meeting of 1791 – when Billingsley himself was in the chair – he was awarded the premium of 10 guineas for the 'Best and Cleanest Crop of Corn, and his Farm in Best Order'.

Billingsley then turned his attention to sheep. Sheep breeding was improving rapidly at this time and there was an on-going debate as to the most profitable breed. Both sheep and profitability being subjects which naturally interested Billingsley, when an opportunity presented itself he decided to act. At the annual meeting of 1791 he was able to persuade two farmers 'of eminent rank in the breeding line', who held strong opinions on different breeds, to 'submit to a fair and unbiassed experiment'. One of these men preferred the new Leicester, the other the Cotswold. In 1792 he carried out an experiment, resulting in an article in volume VII of the *Letters and Papers*: 'A Particular Return of an Experiment made in Sheep-Feeding' (not published until 1795).[3]

Billingsley's idea was that 'five two-tooth wedders' (castrated rams of about two years old) from each breed would be sent to his farm. Other farmers joined the experiment, so that six breeds of sheep with variable characteristics were eventually chosen. They would be kept together for a whole year, under his 'care and superintendence', folded (temporarily confined) with extra feeding each night, and all treated alike. They would be weighed monthly and slaughtered at the society's next annual meeting (December 1792), when 'public testimony should be given of the merit or demerit of each'.

Billingsley claims to have been 'uninfluenced and unbiassed', awaiting the end of the year's outcome 'with anxiety'. The results were very detailed but difficult to interpret and less useful than either he or his audience (then or now) might have hoped. In retrospect this can be seen as due at least in part to a failure to establish clearly defined aims and criteria when setting up the experiment (for example, no statement as to what was meant by 'merit'), and also due to the ranking being subjective (no criteria given). Billingsley disagreed with the rankings given by the judges and gave his own, as shown in the table below. Both the judges and Billingsley were surprised by the poor performance of the new Leicester (Dishley) sheep; this result was excused by the sheep being inferior specimens of the breed. He gave only partial details as to the sheep's feed, though weights were recorded regularly. It is also worth noting that at a time when wool was at least as important as meat it had not been determined whether wool or meat should take precedence or both be judged equally. As he himself points out, the judges valued long-staple wool over short – but which should be seen as more valuable had not been pre-determined. And, while he calls the paper an 'Experiment in Sheep-Feeding', the top of his results chart gives the heading as an 'Experiment on Fatting Sheep'. Neither description would seem to fully represent an attempt to establish the 'profitability' of each breed of sheep, nor their 'merit', though the former was his stated intention. Overall, it was not what we would think of today as a 'scientific' experiment – admittedly not really to be expected for that time.

Breed [5 sheep for each]	Ranking by Judges	Ranking by Billingsley
New Leicester	3	(4)
Cotswold	2	(4)
Southdown (1 died in April)	1	(1)
Wiltshire	5	6
Dorsetshire	6	3
Mendip	4	(1)

Table of Rankings, by independent Judges and by Billingsley himself [it appears 1 is high]

While it was a brave attempt to decide the argument between the breeds, Billingsley himself was unhappy with the results. In the event they seem to have been ignored by the farmers of the area: the new Leicester remained especially favoured, while the Mendip – which Billingsley said in the report he would 'choose above all others' – very soon became extinct.

On this occasion he did write up his experiment in time, although there are signs that the paper was rushed, a certain amount of inaccuracy contributing to a general lack of faith in the results. It was not his best paper. It was an extremely busy period in his life, but the subject had obviously interested him sufficiently for him to cram it into his schedule. At the annual meeting of 1793 he was awarded 10 guineas for the experiment and a further 10 guineas for his 'trouble and expense' in undertaking it, plus it was set for publication in the next edition of the society's proceedings.

Turning to more general society business, Billingsley's influence can be seen in the premiums listed in 1792. For example, in those for the double-furrow plough, which name him in the rules for the premium – offering the cost of a new plough as one prize and for another premium a plate worth 5 guineas for an improvement on the existing design. It was very unusual for the society to name an individual in the rules for a premium, but Billingsley was considered to be crucial to that plough's introduction. Today it is not thought particularly important in comparison with the rest of his expertise. In the section on trees a premium was offered for planting groups of useful trees: walnut, chestnut and beech instead of firs. Other premiums include several for fruit trees. At about this time Billingsley was improving his own property at Ashwick Grove, adding a stand of beech (still in existence) and a large number of fruit trees (disappeared: it is often said that apples do not grow well on Mendip).* It was common for the society to have a preoccupation with one or more subjects at a particular time, trees obviously here being one, so that several premiums would be offered on the one subject.

Meanwhile, the long saga of the merino sheep had begun. In the summer of 1792 Sir Joseph Banks had informed the society that the King offered them the use of two Spanish rams, to cross with members' own sheep – the King of course being George III, also known as 'Farmer George' for his obsession with agriculture. The Spanish sheep (much favoured by the King) produced the finest merino wool. The idea was that English wool would be significantly improved by breeding with the merino. One ram was to go to the 'Earl of Ailsebury' [sic] and it was agreed the other should go to John

---

* Some time later Billingsley introduced the 'Court de Wick' variety of apple to the society. It was well-received, but is almost forgotten today.

Billingsley. The Earl should choose between the old ram – which could serve 80 ewes, and the young ram – 50 ewes. Billingsley was given the old ram. Almost immediately Billingsley signalled his dissatisfaction: 'the Gift appears from age and infirmity to be totally incapable of serving the ewes'. The intention had been that following a ballot of society members, those who were successful could take their ewes to the rams, with strictly controlled numbers, breeds and duration. The lambs produced were to be examined at a meeting of the society the next year. This could now only involve the young ram, making the experiment much less useful. Its performance was not recorded. In June 1794 Billingsley reported that the old Spanish ram was dead. It took the society a whole year before the Earl was asked to report the death to the King. The society hoped to be awarded two replacements, which arrived in due course.

By the mid-1790s Billingsley was focussing more on Enclosures, as a means to reclaim 'waste' land. This subject was also occupying the minds of the rest of the society, which added Billingsley, acknowledged for his expertise in that area, to its committee for Enclosures (at that time he was already busy on the committees for the revision of tithes and for the poor law, as well as agriculture and planting). He was chosen to write the wording of the society's suggested 'General Inclosure Act' (see chapter 8). However, this early attempt by the society was not successful: the Board of Agriculture considered an act for Enclosure too difficult to pass at that time and declined to take it forward. It took until 1801 before the next General Enclosure Act was passed by parliament, not directly through the efforts of the Bath and West.

At some point prior to 1796 Billingsley must have visited Ireland, though no record of the journey survives. We know nothing of it apart from his passing mention in the *General View..* and the following in a letter of 1796, headed 'Foreign Agriculture, or an Essay on the comparable advantage of oxen for tillage in comparison with horses, written by the Chevalier de Monroy'. According to de Monroy, Billingsley had reported that he had

> observed on Lord Shannon's farm in Ireland… three ploughs at work on a strong soil, each drawn by a pair of oxen abreast… ploughed an acre a day without a driver. He further says his Lordship has informed him that two moderate sized oxen had drawn home three tons of sheaves of wheat two miles with no apparent extraordinary exhaustion.[4]

In *The Annals of Agriculture* Young commented that in the above letter 'the comparison terminates decidedly in favour of oxen'.[5] However, the editor of the *Monthly Review* of September 1796 claims that those reading the letter had 'reaped little instruction either from the essay itself or the notes with it, … [we] cannot see anyone being persuaded to use oxen in place of horses' and that 'the subject may be said to be exhausted'.⁶ From comments here and elsewhere it is clear that even by then Billingsley was definitely in the minority in his championship of oxen, though he was still reluctant to give up the cause.

Returning to sheep, in 1796 Billingsley won the premium for the finest ram's fleece, a plate worth 5 guineas. The sheep of his main competitor, Dr Parry, were 'in a state of foulness', needing washing.* Following this win Billingsley seems to have stepped into the background as regards sheep – perhaps he lost interest in them – while the Marquess of Bath (via his steward, Thomas Davis) and Dr Parry became the most active in the merino wool experiments. The Marquess soon carried out and published the results of 'An Experiment on Sheep', using the same six breeds Billingsley had used two years previously. The subject remained a popular one for the next several years: in the 1796 volume of the *Annals of Agriculture* there were reports on two further experiments of a very similar nature.

Ultimately, the results of introducing Spanish rams were not as good as expected. However, in 1802 the King, still in hope of improvement in wool, donated yet another ram to the society, again with mediocre results. It is probable that the breed is not well adapted to the English climate so never reached its potential here. It is now understood that, in addition to genetics, wool is also affected by climate and pasture. Changes in fashion towards the carcass and away from wool, together with the effects of Enclosure, as Parry argued in his 1807 book on sheep, meant that with time the Spanish influence became redundant.†

In the meantime Billingsley was busy pursuing his agricultural interests in other spheres. In addition to the awards he won from the Bath and West Society, he also won a Royal Society for the Encouragements of Arts (RSA) prize in 1797 for 'An Essay on Improving Land Lying Waste'.⁷ Given the choice of either the gold medal or a silver medal and twenty guineas he chose the latter – ever practical. The essay described his purchase of 124 acres of

---

* Dr Caleb Hillier Parry (1755-1822), a prominent Bath physician and an enthusiastic fellow member of the society. He became increasingly involved with sheep and was awarded the society's most prestigious medal for this work. His connections included Sir Joseph Banks, Edward Jenner; William Herschel; and he was brother-in-law to the society's president, Sir Benjamin Hobhouse.
† For a nation whose wealth is said to have been founded on wool, relatively recently it was often regarded as worth almost nothing – some farmers burned it after shearing or paid to have it taken away. It is now again seen as having some value, with efforts to diversify its use, for example as insulation or as slug deterrent.

waste land on Mendip (at Green Ore), including as usual a meticulous account of the process he went through over two years to improve the land, together with his exact expenditure and its consequent value (this important essay is considered in more detail in chapter 8).

During these years the Bath and West Society was extremely active and successful in undertaking trials and in advertising the results of improvements in farming. But the relatively new Board of Agriculture evidently did not adequately realize the extent of their work. In 1799 the board sent a circular to agricultural societies recommending that to help improve agriculture 40 members should give a subscription of 5s per annum for the best cow and calf, best ten ewes and two days ploughing. A reply came back from the society to the effect that they already did something just as useful – and, though unsaid, had been doing that and more for over twenty years. The Bath and West was certainly ahead of the game. And Billingsley had won premiums in a large number of the classes they offered.

The year 1800 marked more than a new century for the society. William Matthews resigned as secretary. He had first threatened resignation at least ten years earlier, citing an excessive workload and the need to be in the society's rooms for long hours, but had been persuaded to stay by arrangements which gave him more freedom to carry out his work in his seed business. In 1800 he really left the secretaryship. This was a loss: he had served the society well in his ten years in the role. However, he remained an active member and stayed at Hetling House. Billingsley was on the committee tasked with appointing a new secretary – Nehemiah Bartley. Like Matthews, Bartley ran a business including provision of seed, but Bartley's Nursery (on the turnpike road near Bristol) was at a distance from his duties in Bath. Unfortunately, Bartley's appointment proved to be a mistake: within a short period he was in dereliction of his duties. There followed a dispute which rumbled on for several years, badly affecting the smooth running of the society, and co-incidentally taking up a good deal of Billingsley's valuable time.

In other respects the society was in a healthy position. It had a surprisingly large membership, catering to a very wide variety of agricultural interests, as shown by the rules and orders book of 1801. Billingsley himself remained active, but whereas in 1792 he served on the standing committees of agriculture and planting, of manufacture and commerce, and of books, plus he had recently also been on those for tithes and for Enclosures, at some time before 1801 he had relinquished all these memberships. He now moved more into judging roles and serving on committees for awarding premiums for livestock and cattle, for crops, for turnips, and for 'cloth and Cassimere' (this was not cashmere, but a thin twilled cloth). He was also on the committee for revising rules and orders, which was especially onerous, revision being undertaken every three years (in 1801 this committee met sixteen times). And, of course, he remained a vice-president, which involved chairing meetings, sometimes including the important and lengthy annual meeting.

For Billingsley, in terms of competitions the previous century had ended with him (unusually) coming only second in 1798 in another of the society's annual ploughing matches, winning a 4 guinea plate. His first place wins resumed in the new century with success in 1800, for the best cow and offspring (5 guineas); 1802, for a fat chinese cross pig (5 guineas – the only competition he won involving pigs); 1804, for ploughing with neat cattle (another 5 guineas); first prize at his last entry to the trial of ploughs (6 guineas); and 1805, for the best pair of working oxen (another 5 guineas).[8] Following these wins he seems to have stopped entering stock and ploughing competitions, but was still very much involved in agriculture and the society. While for many years previously he had appeared only at the annual meeting, his attendance became rather more regular during the years 1803-8 while he was living in Bath.

In February 1802 a serious event for the society occurred: the premature death of its president, the first Duke of Bedford. While etiquette required that the president of a society of this nature was a member of the nobility many were not effective in the role – unfortunately this had been so for the society with its previous presidents. Bedford was an exception. He had been extremely useful so his loss was much regretted. It was immediately decided to strike a medal in his memory: the Bedfordian Medal, value 20 guineas, to be awarded for outstanding contribution to agriculture. This medal became the society's most prestigious award, not given every year and for a long time reserved for some very special achievement.

Designs for the medal proved difficult, the first, prepared by a Miss Fanshawe, being rejected.* The commission was then given to Mr Milton, but the design was not finally agreed until 1805. The Duke's brother, the second Duke of Bedford, replaced him as president. This, however, did not prove to be such a happy arrangement: he had resigned by 1805. He was succeeded in turn by Sir Benjamin Hobhouse at the end of that year, an active and successful long-term president.

In 1803 a new committee of superintendence was formed, to be elected

---

* Rejection was due among other things to the legs of the sheep being too long and the bull 'lacking any vestige of a penis'. She was advised as to where she might seek information as to a bull's anatomy.

annually, with Billingsley as one of its five members.[10] It was effectively a management committee, made up of senior members of the society. Members were required to live within the environs of the city of Bath, which suggests frequent meetings both formally and informally.

While membership of the Bath and West Society was in many ways of great benefit to Billingsley, he definitely benefitted the society in return, not only through his agricultural expertise but also in ways unrelated to agriculture itself. One such was the 'Bartley Affair'. This was one of the first issues the new superintendence committee tackled. Bartley was variously accused of misuse of funds and incompetence. There is no evidence that he ever did anything actually fraudulent: it seems he was simply inefficient. He failed to attend agreed meetings and was difficult to pin down as to his whereabouts,

Two pieces of Silverware awarded as Prizes by the Society to John Billingsley Esq L: for the Best Pair of Working Oxen, 1805; R: for the Best Cow and Offspring, 1800[*9]

* On the original photograph the words 'Bath and West' can just be discerned on the left-hand item, and 'Bath Society to Mr J° Bill...' on the right-hand item (photographer unknown). The crest on the left is not that used by the society today; it remains unidentified. Mrs W Moger, a direct descendent of Billingsley's, held these and other items of his silverware at her death in 1963, bequeathing these two pieces to members of her husband's family, with whom they are believed to remain today.

prompting increasing outrage among sections of the membership. From the evidence available, by modern standards Bartley would probably be thought to have been suffering from mental health issues. Billingsley was a main actor in sorting out the situation, often (but not always), chairing the relevant meetings, and (with the help of Matthews) untangling the accounts. Whether Billingsley's service in that role was requested or volunteered is not clear, but it is typical of him to be in that position (he had shown the calm-but-firm negotiator side of his personality early in his professional life, during the Shepton wool riots of 1776).

The Bath and West Society's Bedfordian Medal produced from 1804 by Thomas Ottley of Birmingham[11]

Bartley was not good at administration and probably overworked, especially as he was also running his own business at a distance from Bath. Unfortunately for him the superintending committee was very alert to matters financial, and Bartley's records tended to be what they regarded as slovenly. The problem had been simmering from 1802, then in September 1805 it came to a head. Matthews checked Bartley's accounts and found them wanting. A meeting chaired by Billingsley took into consideration Mr Bartley's version of accounts, complete with the latter's claim to a surcharge. Mr Matthews produced his own statement 'to shew the error of [Bartley] claiming to make a surcharge of £42 18s, but Mr B[artley] had rejected it, and treated the writer with scurrilous language'. It was 'unanimously thought proper Mr B[artley] should not attempt to insert it into his account at the upcoming General Meeting'. Not only was the problem causing a dent in the society's finances and smooth running, but also important, it was disturbing camaraderie, prompting arguments among the members.

Things continued to go downhill. Billingsley, and the committee as a whole, was firm and dogged. The annual meeting decreed that Bartley's services as secretary should cease on 21 December, when he was required to 'quit the premises after delivering up to the custody of Mr Matthews all papers and property…belonging to the Society'. Bartley was asked on several occasions to state precisely the date and time when he would be ready to meet the parties for examination and final adjustment of accounts, but such requests were repeatedly frustrated. The committee's minutes accuse Bartley of 'evasive behaviour' and 'intentional disrespect' by leaving gentlemen waiting in vain for him to attend agreed meetings.

Billingsley wrote seemingly endless letters, almost all followed by no-shows. After a meeting in February, at which Bartley did appear, the committee subsequently thanked Billingsley, as chair, for his 'candour, attention and ability in conducting the business of the day'. Finally, in late May an extremely patient Billingsley wrote yet again, still very politely, expressing the committee's great displeasure, and that failure to make a final account would result in the matter becoming public by being laid before a general meeting. This 'very unpleasant and troublesome affair' was finally laid to rest in June 1806, when the committee determined that Mr Bartley, who after the society had been very generous with respect to money awarded to him, still owed them £22 16s 2d – probably never paid.

It took until June 1806 before the society was finally rid of the unfortunate Mr Bartley. Following this debacle the new secretary Robert Ricards (or Rickards) might well have been on his guard. Ricards was a print seller and stationer, chosen by ballot and appointed in 1805, an improvement on Bartley from the beginning.

Volume X of the society's *Letters and Papers* appeared in the same year, delayed, one is led to believe, by the inattentiveness of the previous secretary and edited in his place by Mr Matthews. Three papers in this volume are particularly relevant to Billingsley. First, an account of the inspection of Mr White Parsons' farm in West Camel, for which Parsons won a premium. The judging team of four was headed by Billingsley. The team had obviously not just observed the crops and methods used but also questioned Mr Parsons closely on his views and approach to farming. For sheep 'Mr Parsons was a decided enemy to the folding system', which was then in vogue (and favoured by Billingsley), but a 'warm advocate of best foreign blood', then generally seen as necessary for successful breeding. On the negative side 'Mr Parsons had not kept a regular account' of costs (a fact liable to produce at least private censure by Billingsley), which meant that 'it was not possible to determine the expense of the project'. In Billingsley's eyes this at least will have served to greatly diminish its usefulness, though nothing is made of it in the report, which merely observes that the committee felt the land was now probably worth £500 to £600 a year. On the positive side, it notes that Parsons believed that 'ashes are much better than marl for strong land and cheaper than lime': according to the team this relatively novel approach deserved more attention. Given that Parsons's methods in general were so contrary to Billingsley's own the latter was unbiased and generous in his report.

The second, more important, paper in volume X is Billingsley's own 'Remarks on the Utility of the Bath and West of England Society, with an Account of the Progress of Improvement in the County of Somerset'.[12] This article could well have been commissioned by the society: the content reads as if it may have been first given as a speech to them. However, although the title looks promising, the first part of the essay would have been more useful had it not tended towards assertion, lacking Billingsley's usual facts and figures. The later sections are more effective.

He begins by stating that, among similar societies that have sprung up in the last 50 years, the Bath and West Society 'has been honoured with peculiar marks of distinction and approbation' and that such a conclusion seems to be supported by facts. He then goes on to say that there are many people, such as 'conceited farmers, who hold this and all other similar establishments in utter contempt, and seize every occasion to manifest their disapprobation in dislike'. This section is followed by a potentially useful attempt to prove the excellence of the society 'from its effects'. He observes that if Somerset has made more rapid strides in improvement during the last 30 years than other counties, then this might be due to the 'vigour and energy' produced by the society, and by the frequent association and discussion among its members. Again, the inclusion of facts and figures would have been more persuasive. Although Billingsley does not actually make the point it does seem to be true that most agricultural societies of the time were more social clubs than the Bath and West, which was renowned for its positive programme of premiums, trials and experiments. Many improvements had been undertaken which 'if not exactly under the influence of the society, have proceeded under its notice, and by persons who have taken a conspicuous part in its direction'.

He then turns to the area and topics he knows best – and includes the facts and figures missing hitherto. Billingsley says that on the Mendip Hills upwards of 20,000 acres of land have been Enclosed since the establishment of the society, its value improved from 3s 6d to at least 15s per acre; the amount of corn, particularly oats has increased immensely, without this additional produce grain would have been sold at an enormous price. This may be considered as more important, he feels, than the £12,000 per annum advance in the annual rent of the county. And, 'one of the principal promotors of this grand scheme of improvement has received as many, if not more, premiums for crops, than any other member of the Society' (no name given – presumably himself).

In addition, he says, 30,000 acres of marsh land have been drained and improved by at least 20s per acre; besides 10,000 acres enclosed in other areas of the county meaning an advance in rental of at least £60,000 per annum. All of which improvements have rendered the county one of the most productive in the kingdom, enabling it to send thousands of fat oxen and immense quantities of cheese to markets. He goes on to say that:

> while these great things have been doing in this county, and whilst nearly 60 bills of inclosure have been carried to parliament, the adjoining county of Devon, abounding with wastes, has done, comparatively speaking, but little. This, therefore, is a presumptive argument, that the establishment of this Society has in some degree contributed and excited such grand improvements.

Billingsley claims, then, that the society itself is the catalyst to improvement. Subsequent commentators have seen this as being the case, with the Bath and West Society having an excellent reputation as a facilitator in agricultural improvement, especially during this period, the first 30 years of its life.

Immediately following Billingsley's paper a section by the editor (Matthews) explains that Billingsley had intended to 'follow up these remarks with further observations on different kinds of improvements, prices of provisions, etc; but that being taken ill, and his health suffering considerably, he thought himself obliged to lessen his cares, and to recommend an enlargement on those topics to some other person.' (This, incidentally, is an early intimation of Billingsley's health issues, which would soon peak, then seem resolved, before becoming serious again – see later). The editor continues by making the point that 'it is no more than justice…to give to the credit of Mr Billingsley himself a very large share of those improvements'. The editor cites a Wiltshire gentleman, who was tasked with judging Mr Billingsley's farm for a premium, 'who evidently came under *prejudice* against the value of the improvement'. Having, however, surveyed the farm, he stood on a raised spot and declared that 'in justice a *statue* ought to be erected on that spot to the fame and memory of the improver'. Leaving aside the eighteenth century hyperbole this is a deserved compliment.

The third relevant paper in this volume returns to the use of oxen – a debate thought to have been decided in 1796 in favour of horses, yet evidently not quite dropped by some of its more stubborn protagonists. Here, Lord Somerville gives an account of his use of 'Oxen and Neat Cattle in Husbandry'.[13] He then waives his own opportunity of a premium in favour of Billingsley, whose paper on the same subject follows. In this, 'A Practical Statement on the foregoing subject with Claim to a Premium', Billingsley states that on a farm of 800 acres, with a team of six oxen, 385 acres were ploughed and 291 harrowed in eleven months. In his opinion, though oxen 'will not answer on every situation', on level ground without stones, oxen and a double plough are preferable to horses. But further discussion on the subject does not seem to have been reignited. Despite giving facts and figures in support of his view, this is one battle that Billingsley emphatically lost.

In 1806 the society had offered the Bedfordian Medal for the best essay on the cultivation of common land. Billingsley was successful in winning the medal – the highest honour the society could bestow – for his 'Essay on the Cultivation of Waste Lands'.*[13] The 1806 the minutes of the annual meeting record:

> Resolved, that the Bedfordian Gold Medal be given to John Billingsley Esq, with the thanks of this Society for the fresh instance of zeal and ability in adding to the stock of practical agricultural knowledge so eminently manifested in the Essay now presented to the Society.

The full paper was published in volume XI of the society's *Letters and Papers* in the following year.

By this time Billingsley had been working for about 30 years on bringing so-called 'waste land' into cultivation. He was widely regarded as an expert in the field, his expertise coming from his own farming practice and his work as an Enclosure commissioner (see chapter 8).† Despite the *General View..* being by far the best known of Billingsley's publications, in many ways 'Waste Lands' can be seen as of almost equal importance, at least in practical farming terms. It is the most detailed published description of his farming methods and practice to be found in any of his publications. 'Waste Lands' may have a different emphasis, but has many of the same positive attributes to recommend it as the *General View..*, for example being written at a time of agricultural

---

* Called 'Waste Lands' hereafter.
† He will also have benefited in a less obvious way from his membership of the Bath and West and its connections.

change and containing innovative ideas. But while every county has the benefit of a *General View..,* Somerset is probably the only county with a primer of this calibre in how to achieve significant advance in the agriculture of the time. It is the clearest demonstration that he was ahead of his time in farming practice. Recognising its value and that its content is relatively unknown today, 'Waste Lands' is considered in detail in chapter 15. This seeks to redress the imbalance in attention by giving a synopsis of the essay together with discussion of its relevance to current agricultural thinking and practice.

We do not know whether Billingsley was still experimenting and supervising agricultural practice on his own land at the beginning of the new century, but he entered for no more premiums and wrote no more papers for the society. His intermittent ill health at that time would make it easy to understand any reduction in enthusiasm. However, he continued to be very active in Enclosure work, which would have been extremely time and energy consuming. He also attended more meetings at the society in the years 1806-10 than had been his usual practice, and took an even bigger part in its running than previously, serving on the committees for agriculture, 'cyder' [sic], judging for the cloth and servitude premiums, plus being very involved in the superintendence committee. In today's terms he can be seen as having gradually left his role as youthful activist and taken on that of management and advice as a society elder.

The change in his pattern of meeting attendance was probably at least partly because he was resident in Bath for much of this period, so would not have had to travel far. In 1810 the committee reported 'considerable slackness of attendance at committee meetings, due to great national anxieties'. Billingsley did not approve inactivity, so suggested that vice-presidents who had not attended for two years should not be re-elected. He was present at the annual meeting in December 1810 (though by then living back at Ashwick Grove). He had not entered the last trial of ploughs, but the report on it must have been discouraging for him: there was said to have been a 'general waste of strength and expense' in the methods used. The meeting also suggested ways the 'numerous poor, liable to be out of employ, might be helped'. There is no record of Billingsley's response to that. A meeting in February 1811 was his last for the society. He died unexpectedly in the September.

8
Enclosures

BILLINGSLEY WAS AN enthusiastic champion of Enclosure. By the start of the 1790s he was heavily involved in the process. In his list of 'Hints for Improvement' (at the end of the *General View..*), the first item he gave – and for him probably the most important – is: 'Inclose and cultivate all waste lands susceptible of improvement and divide and inclose all the common fields'.[1]

At its simplest, enclosure is the process of converting what had been open land into plots with some boundary – a hedge, stone wall or other partition. This allows change in the use of the land and is often related to change in its ownership. It also results in change to the land's accessibility. In earlier times some open land had been cultivated as shared fields, which were gradually enclosed. Other land had been open common land (mostly rough unmanaged pasture) on which local inhabitants held 'rights of common' for grazing stock and for collecting wood, furze and so on for fuel. These rights went back into the mists of time. Since then land has been gradually taken out of collective ownership and passed into private hands. The boundaries serve to show both the extent of the privately owned land, and to keep animals – and humans – in or out.

There were various reasons behind Enclosure (that is enclosure by parliamentary act). On the one hand a landowner might simply Enclose or sell common land for his own gain, or to form a parkland setting for his great house. On the other hand the drive to Enclose might be for more or less pure economic and/or agricultural reasons: 'improvement' of the land, leading to greater yields and greater profit.* But, importantly, Enclosure denied the use of the common to those who had previously held rights of common: most of the later eighteenth century parliamentary acts explicitly stated that 'Right of Common [was] to be extinguished'. Overall, the intentions behind Enclosure were diverse and not always clear. In many areas it had devastating social effects, especially affecting the poor. However, there are some ways in which

---

* Improvement was seen as enhancement of the capacity of the land to produce a return through agricultural methods, often through Enclosure.

Enclosure can be seen as ultimately for the greater good: for example improved productivity at a time of shortage of essential goods. Different areas of the country were affected to different degrees and in different ways: Mendip in particular was not typical.

While some enclosure had taken place for centuries, the first Enclosure by parliamentary act occurred in 1604. From then on acts enabled the process, which increased markedly in the eighteenth century. In the southwest Enclosure was encouraged by organisations such as the Bath and West Society and by individual advocates including Billingsley. As by his day the great majority of land in Somerset had already been enclosed, any significant remaining areas (the uplands, including Mendip, and low-lying areas prone to flooding, such as the Levels), was land hitherto regarded as unattractive for agriculture.* These areas were decidedly less prosperous than nearby regions. The 1742 Proportion Roll, for example, shows the proportion of tax payable based on the wealth of the area: it indicates that Mendip had to pay only half the amount as lands immediately to its north and south.[2]

During the second part of the eighteenth century England entered a period of economic difficulties. Rising prices (especially for wheat), plus unrest in Europe and threat of war, or actual war, meant it became increasingly apparent that the country needed to expand its food production. Those advocating Enclosure for economic and agricultural reasons, including Billingsley, maintained that in Somerset the most effective way to achieve this was by bringing common land into use. It was believed that common land, or 'waste', was just that: wasteful, less productive than it should be. Arthur Young claimed that as 'many judges have questioned whether all the commons in the kingdom are worth a groat to the public, the enclosure, division, and cultivation of them in any degree, must be a great advantage indeed'.[3]

They believed that any land that *could* be improved *should* be. The waste land Enclosed would then be distributed between existing landowners and tenants, increasing their holdings in order to increase productivity (and financial gain). Billingsley explained the legal process thus:

> There are but two modes of inclosing commons. First, by unanimous consent of the parties claiming rights, who delegate power to Commissioners, chosen

---

* While historically the majority of Mendip upland had long been the 'Forest of Mendip' (royal hunting ground) this use had been rescinded by medieval times. So, prior to the eighteenth century Enclosures Mendip commoners were using the open land as usual for commons, for grazing their stock and collecting fuel. Large areas of Mendip were also used for mining.

by themselves, to ascertain their validity and divide accordingly under covenant and agreements properly drawn and executed for the purpose. Or, secondly, by Act of parliament obtained by the petition of a certain proportion of the commoners both in number and value; whereby a minority, sanctioned only by ignorance, prejudice, or selfishness, is precluded from defeating the ends of private advantage and public utility.*⁴

The second way he mentions (regrettably expressed) was by far the more usual. However, there are some local examples of the first, such as Lord Fortescue (lord of the manor of Ashwick and owner of almost all the parish) effectively Enclosing Ashwick down, including the great majority of its common land, all owned by him, in the early 1760s.⁵

If using an act the process of Enclosure was generally set in motion by the major landowner(s) of the area in question, most often a parish. In outline, an initial survey was undertaken to demonstrate the extent of the land and identify its main owners, following which an application was made to parliament for an act – each Enclosure needing its own act until 1801. During Billingsley's time there were three general Enclosure acts (1773 with several amendments, 1796 and 1801) which gradually made the legal process easier and less expensive. Commissioners (usually three) were appointed under the act. Arthur Young described the extreme authority given them:

> The power granted by most of these acts to Commissioners, is an extraordinary circumstance in the History of Enclosing. They are a sort of despotic monarch, into whose hands the property of a parish is invested, to recast and distribute it at their pleasure among the proprietors; in many cases without appeal.⁶

Commissioners were responsible for overseeing the division of land in proportion to the existing holdings of the various owners. They also oversaw the financial side; the outlay including surveyor's fees; the identification and eventual construction of necessary roads; fences and walls; drainage ditches; gates, bridges and so on. Commissioners were obviously chosen as trustworthy,

---

* His description of the 'minority' here immediately signals Billingsley's social prejudices. While the minority sometimes included landed gentry who wished their 'view' to remain unchanged (see later for an example), Billingsley is probably thinking more of the landless majority here, who feared losing their rights of common. Most reports on Enclosure recently have tended to concentrate more on the social repercussions than on the agricultural effects. While the main focus of this book is very much on the latter, as Enclosure has been such a contentious issue the former is also considered to some extent.

it being something of a compliment to be appointed. They were frequently local landowners or agents, so were familiar with the type of land. The commissioners appointed surveyors who were often very skilled, the resulting maps beautifully drawn, many being the first detailed map of the area up to that date.

The Enclosures Billingsley was active in were centred on his home locality of Mendip and the Levels, although his influence was much wider (the Levels' acts are covered in chapter 12, together with the drainage involved). It was clear at the time that Mendip was not ideal for agriculture. The central plateau ranges in height from about 600ft (183m) to over 1000ft (up to 325m). At many places along the edges the land falls away very steeply. There was – and is – a very low population density, any habitation clustered round the edges of the area. There was only one small village on the whole plateau. The soils are shallow, often poor and stony and/or marshy. The land is exposed to south-westerly winds, including gales from the Atlantic, yet despite the very high rainfall there is a noticeable lack of free surface water due to percolation through the limestone. Trees were then few, stunted or absent. Until shortly before this period a substantial area of Mendip had been mined for the lead and other ore found there. By Billingsley's day this work was beginning to tail off, though spoil from mining littered the landscape and caused on-going pollution. In addition, the weather affected what stock was kept or crops grown. None of this seems to have dampened enthusiasm.

Billingsley contributed to the Enclosure movement in three main ways: first, as a commissioner for a number of parliamentary acts; second, by bringing the newly Enclosed land into cultivation (no doubt watched with interest by those in his network of contacts); and third, through his published work, including the *General View..* and other papers which detailed his pioneering methods. All three ways helped to disseminate his methods and views, his acknowledged experience meaning he wielded a significant degree of influence, especially locally. He was given a large part of the credit by his contemporaries for Mendip's Enclosure, as well as for the Levels' drainage and Enclosure. They were impressed by his methods and results. He also, of course, derived personal benefits from the process, financial and otherwise.

Beginning with Billingsley's role as a commissioner, his first appointment was in 1789 (West Harptree), his involvement continuing until his death – at which time he was still actively working as a commissioner on at least four acts. He served on seven of the fifteen Mendip acts in the period 1779-1800 (more than any other commissioner for Mendip acts).[7] He was also a commissioner for the act for Cheddar, Priddy and Rodney Stoke in 1811 (but died within

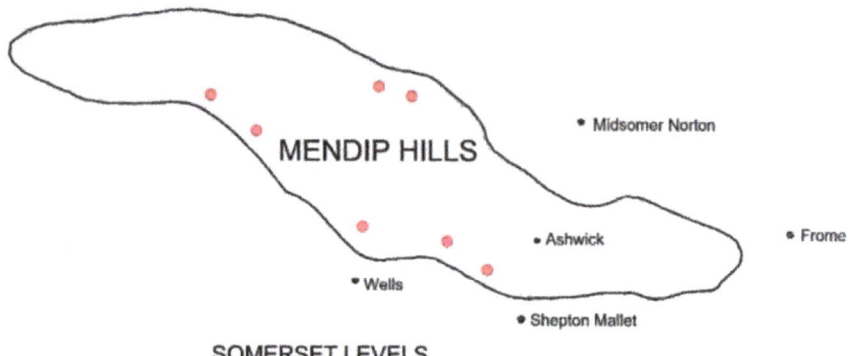

Sketch Map of the Mendip Hills
Enclosures for which Billingsley was a commissioner are marked in red

months of it starting so did not complete the process); and for at least three acts just outside Somerset not complete at his death (one in Wiltshire and two in Dorset; for the latter two he was sole commissioner). Plus, although not a commissioner for the Chewton Mendip act (1800) he played a crucial part in steering it through in the face of difficult opposition, producing some 1,500 acres of land for Earl Waldegrave. He then 'reclaimed' the land for the Earl. Finally, in the *General View.*. he anticipated the usefulness of an Enclosure act for Exmoor, with a description of the way the improvement could be undertaken, twenty years before others took it forward, using many of his ideas. He had no hand in the act itself, which was not passed until well after his death.[8]

The table below gives information as to the extent of the Enclosures which Billingsley was involved in, together with some idea of the time the process would have taken (though some of the work will have been done before the date on which an act was passed). His activities as a commissioner peaked in the decade between 1790 and 1800, with six of the seven acts he was part of being passed during those years. This decade was also the climax of the Mendip Enclosures in general, not just the number of acts but also the acreage Enclosed. Commissioners were chosen at the time by the landowners themselves, and often from among them (except from those with major holdings within the scope of the act). This demonstrates the trust in which Billingsley was held by his contemporaries, and was also, no doubt, acknowledgement of his expertise. He seems to have been much in demand. As a commissioner, he oversaw the necessary financial arrangements and legal and other paperwork (including personal agreements) for each Enclosure. He was also very active in overseeing the practical side, visiting the fields and work on the land as it was undertaken. It represented a huge amount of effort.

For each act, records also show the outcome in terms of who land was allotted to, how much land each recipient was granted, and which plot(s) they gained. Allotments were generally based on the quantity of land already held (a 'to the rich shall be given' approach), divided proportionally. Those without land but with commoners' rights were therefore unlucky, although tenants of even small cottages might acquire small allotments or additional land around their dwellings. Some land was often, but not always, reserved 'for the Benefit of the Poor of the Parish', any proceeds contributing to funds for poor relief. Commissioners were required by the acts to manage costs and empowered if necessary to sell the portion of land they thought was needed to meet them. For Mendip Enclosures in the 1790s the fees averaged £1,336 per act.[9] This was a period of inflation; costs for Mendip Enclosures in general almost doubled in the period 1770-1810.

Three Mendip acts levied a rate to pay the fees incurred, the rest provided the funds by selling off land at auction. Billingsley was a commissioner on only one of the acts levying a rate. It is not clear from the records how the choice was made, but the two methods had pro's and con's: a levy cost more for the landowners, but resulted in more land available to them; while auctions meant less initial outlay but less land at the end of the process. Most of those who bought land at auction were local farmers and landowners. Billingsley was typical in this respect, purchasing land in several Enclosures. There was,

Name	Date of Act	Date of Award	Extent in Acres
West Harptree	1789	1790	857
Croscombe & Dinder	1792	1793	603
Wells, St Cuthbert Out (East Horrington)	1792	1794	701
Wells, St Cuthbert Out (Mendip)	1793	1795	4,343
East Harptree	1794	1796	1,033
Cheddar	1795	1801	4,400
Cheddar, Priddy & Rodney Stoke	1811	1821 (after his death)	1,100
		Total acres enclosed	12,091*

Table showing Mendip Enclosure acts for which Billingsley was a commissioner*

* As Billingsley's Enclosure work on the Levels included its drainage it is considered in a later chapter.

* Land enclosed for all 15 Mendip Acts in this period totalled 22,873 acres

incidentally, no rule at the time that commissioners could not benefit from land in an act they served on – for example, Billingsley was both a commissioner and bought land at auction at Green Ore on Mendip, considered below. There tended to be a flurry of land trading soon after an Enclosure: allocated plots were often soon sold off (especially small plots, if recipients could not afford the fencing required by law) and apportionments could be exchanged (as Billingsley did following the Chewton Mendip Enclosure). As a commissioner Billingsley was extremely efficient, speedy and effective, despite the large number of acts he served on in parallel.

Turning now to an example illustrating both his second and third contributions to Enclosures (that is, the methods he devised for bringing Enclosed land into cultivation and its associated publication) this concerns the land at Green Ore on Mendip (in which he also benefitted personally from Enclosure). It resulted in his prize-winning paper for the Royal Society of Arts and Manufactures (RSA): 'The Improvement of Land Lying Waste', published in 1797.[10] The essay is a case-study of his activity in improvement. The land in question was acquired by Billingsley from the Wells (St Cuthbert Out) Enclosure act for which he was also a commissioner. On 27 June 1794 he attended an auction at the Swan Inn, Wells and was successful in bidding for 124 acres of land at Green Ore.[11] These acres lay west of the Bristol turnpike down as far south as the new Priddy road (a road to be set out by the commissioners of the act). This information is especially useful in that it makes this patch of land one of the few for which we know the exact location of one of his 'experiments'.

Billingsley recorded all expenditure for this project meticulously down to the last penny, as was his custom. The results were published in his RSA paper; the facts behind it endorsed (as required by the RSA) by a member of the city of Wells, Clement Tudway, and the mayor, John Lovell. Land purchase costs at Green Ore totalled £1,115 13s 6d. This payment will have helped to cover the costs of implementing the act. Billingsley began by clearing and levelling the ground which, being on the edge of the lead and calamine district, was littered with 'many large hillocks raised by ancient miners'. He described the land as being 'a gravelly loam; with a limestone rock, at the depth of six or eight inches'. The climate, he said, is 'cold and moist; spring late, and summer short…its altitude is about 1,200 feet above the Bristol Channel'. Being exposed, it may have seemed that high but, in fact, by today's instruments it lies at only 758 feet (231 metres) above sea level.

He undertook a comprehensive programme of cultivation ('improvement'): repeated ploughing, harrowing, rolling, then manuring, weeding and so on; making boundary divisions and hedging (including erecting temporary stone walls and banking to protect the young hedge plants); liming (building and operating a lime kiln, sufficient lime being produced and added); constructing two large pools for stock (Mendip being very short on surface water); purchasing seed for crops (wheat, oats and winter vetches); and constructing farm-yard buildings. As an example of his precise accounting, the costs involved in building the barn are shown below.

		£	s.	d.
	(13) Particulars of cost.			
	158 perch (15 feet) of wall, at 1s. 8d. per perch, quarrying and wheeling the stones included.—N. B. Stone close at hand	13	13	0
	Digging out foundation	1	1	0
	Elm roof, at 21s. per square	16	14	0
Barn.	Doors, linterns, sheaf-holes, partitions, and corn-binn, and lining walls	15	3	0
	Tile, 7,500 plain tile	11	16	3
	Labour, putting on, and laths	9	10	0
	Freestone, for water tabling	1	2	9
	Three-inch elm barn floor	12	0	0
		81	0	0

(14) This stalling is made so as to answer two purposes; namely, a foundation for a corn-mow, and a comfortable residence under for cattle.—Cost as follows:—

Costs of the Barn at Green Ore
as given in Billingsley's essay 'Improving Land Lying Waste' for the RSA[10]

Billingsley's total outlay, including purchase, was £2,453 17s 3d. Against this he set the value of the produce (£784 0s 0d) and of the land now in good order, estimating that he could 'now let the inclosure for £150 per year, or 25s per acre (nearly), the value in Fee, therefore, at 25 years purchase is £3,750 0s 0d, from which we will deduct, for a dwelling house, stable etc £250 0s 0d' – resulting in £3,500 return. Having obtained good initial crops the land was ready for leasing out. These figures are a vindication of his belief that improving the land would result in improved financial gain (although elsewhere he stresses what he considers to be the even greater importance of an increase in crops).[12] He held that improving the soil was a prerequisite to productivity, and that, together with the enclosure itself, any 'improvement' was very advantageous for agriculture – provided the land was then treated respectfully and not overcropped (this last was not always honoured – see later).

The essay gives great detail as to the process he uses to bring the land into cultivation and the costs involved. It not only won the RSA award, but also at least potentially gained more widespread dissemination of his views and methods – the RSA having a more national audience than the Bath and West Society is likely to have given him. It was one of several papers with strikingly similar messages as to the methods he promoted. Other examples include his

'Essay on Waste Lands' (a more general version building on ideas in the above Green Ore project; this is discussed in chapter 15), and 'On the Uselessness of Commons to the Poor' (considered further below). Usually seen as the most important, of course, was his book the *General View.*. which contains long sequences on his methods, practice and views (see chapter 14), though it does not give nearly as much practical information as 'Waste Lands'.

Turning to other aspects of Enclosure, such as the distribution of land following an act, one relevant example is the Shepton Mallet Enclosure Act of 1795.[13] This shows another way Billingsley profited personally through Enclosure. He already held land in the area but was not a commissioner for the act. It involved about 720 acres, though his benefit was modest as he held only a relatively small amount of land in the parish, in several different sections, some freehold and some leasehold.* Most of the land being Enclosed was owned by the Duchy of Cornwall (the crown, lord of the manor) and leased to a large number of individuals. Billingsley was allocated land in proportion to his holdings. On the part of the map shown below he has been awarded plots twenty and twenty-one in the centre, and as Jordan and Billingsley (co-owners of the brewery) plot nineteen just north of them. He is also noted as holding land in the 'Oak-Hill Old Enclosures' immediately north of his two allocated plots, near other land owned by the brewery at that time.† Just visible on the original map, in very faint pencil, alongside plot twenty is written 'Billingsley's Way' – presumably a name already used by locals for the track which led to his land.

The plots are outlined in different colours, visible on the original map, showing the way in which they were legally held: plots twenty and twenty-one (Billingsley) are held leasehold from Fortescue, while plot nineteen (Jordan and Billingsley) is shown as freehold. In addition, as an individual he had an allocation of a further eleven acres or so from land he owned within the manor, but south of the area shown here. Billingsley gained land by allocation from a number of other Enclosures in this way, mainly on Mendip. He is also said to have 'reserved' plots following acts there (considered in chapter 9).[14]

Enclosure has often been denigrated for reducing the population on the affected land. On the Mendip plateau, however, where the population had previously been extremely sparse, it actually *increased* immediately after the

---

* The land was very close to his home at Ashwick Grove, which was in Ashwick parish where he owned a substantial amount of land – but land which would not have counted in this distribution.

† This land had already been enclosed some time ago, not by a parliamentary act but by gradually being taken into cultivation as 'closes', carved out of existing open land.

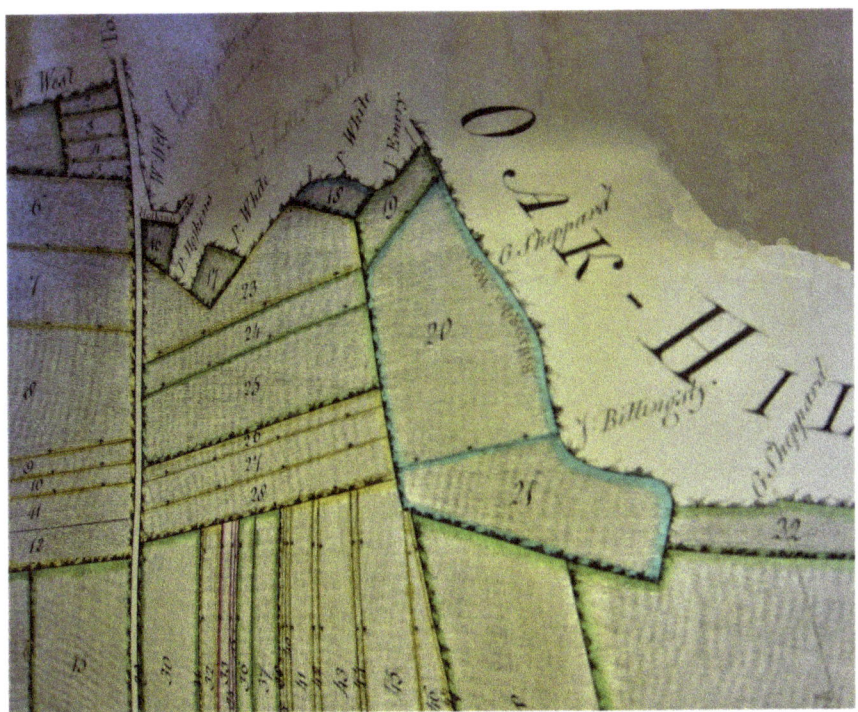

Detail from the map for the Shepton Mallet Enclosure Act of 1785[13]

Enclosures, mainly due to the establishment of farms where there had been large areas with no dwellings at all.[15] However, this does not mean many people on Mendip were not disadvantaged by Enclosure. To the landless who held rights of common this same waste land will have been an important source of sustenance that was now lost. Those with least would have suffered most. Small farmers did not necessarily fare well either.

One common outcome was that small plots – as in the Shepton Mallet example – tended to be consolidated: cottagers and small farmers were often forced to sell out to avoid the cost of Enclosing, benefitting those with greater resources. An example of this can be seen in the manor of Stoke Lane, which Billingsley purchased in 1782 (after its Enclosure); the land bordered his home estate at Ashwick Grove.* Part of the manor had been Enclosed in 1776 under the Doulting and Stoke Lane Enclosure act of that year. According to *The History of Stoke St Michael* 'five farms had gone from Stoke Lane manor by 1783 and the remaining nine farmers had taken over the abandoned land', three of these farms being considerably increased in size.[16,17] Also by then twelve cottages had been built on wasteland along the roadsides and a smithy

---

* Stoke Lane is now known as Stoke St Michael.

had been erected at a crossroads. Perhaps the inhabitants of the new cottages were those who had abandoned the land. On the other hand, by 1783 five new farms had been established on the newly Enclosed land. It is possible that one or more of the original small farmers may have begun again, but equally they may have been new tenants.

A similar effect is reported following the Enclosure of Shepton Mallet in 1785, discussed above. The duchy's agent wrote a report on the estate a few years later: reading between the lines this illustrates some of the less positive effects of Enclosure. As there were nearly 300 allotments most fields were very small, so many of those who had been granted them then sold out, and those who purchased other small fields were able to consolidate them into blocks and then took out leases under the duchy. But crucially, the new leases lacked the original rights of common for that same land. This was one of the processes through which the old rights of common disappeared. In later acts rights of common were totally extinguished. Following this the manor of Shepton formed what the 1790 report described as having been 'a poor barren common' before Enclosure, and now being 'a distinct and very considerable estate which but seven years ago was scarcely worth owning'.[18]

Billingsley appears to have been genuinely convinced, albeit erroneously, that Enclosure did not disadvantage the poor. He does not dismiss the human element entirely, but is so focussed on the agriculture that he has a distorted sense of reality for the landless. In his paper 'On the Uselessness of Commons to the Poor' he claims that the rights of the cottager

> cannot be invaded [by Enclosure], since he stands precisely on the same ground with his more opulent neighbours. I can truly declare that, in all cases which have fallen within my observation, inclosures have meliorated his condition, by exciting a spirit of activity and industry, whereby habits of sloth have been by degrees overcome, and supineness and inactivity have been exchanged for vigour and exertion. No stronger proof can be given of this than the reduction of the poor's rate, in many of those parishes wherein such inclosing has taken place.[19]

Some questionable assumptions are made here, both as to the cottager's habits and the reduction in the poor rate – without any evidence given for either. He goes on to note that when the commons are overstocked then cottagers must leave their stock to be stunted or starved. And while the upland commons, such as Mendip, are grazed by sheep in summer, he says that cottagers are totally unable to feed their stock during the winter, having no alternative pasture or extra food for them. In contrast, those with large areas of land were able to use the commons in the summer, in competition with commoners, and could then accommodate their stock for the winter and arrange feed for them. Like those with land who had invested in their project, cottagers with certain tenancies which came without land but with commoners' rights ('auster' tenants) would receive an allotment and benefitted by not having had to invest, he claimed.* But, as we have seen, they may well have been forced off the land by the cost of Enclosing. And note too that he writes about cottagers, not the landless poor who were in an even worse situation.

On moors (in this context Billingsley means the Levels) cottagers generally turned out a cow or two and some geese – only the latter are profitable, he says. Even in summer the land is frequently flooded, so cattle must be taken off the common and pasture rented. Or, if the season is good, the presence of so much food means the value of the cow falls below what it should be. And cottagers do not rent land for winter feed for them. For 10s or 12s per acre per annum one can rent a common right. Once Enclosed the land is then worth £3 to £20 per acre per annum, he claims, a great improvement for the cottager (again, if he can afford it). In further assertions, Billingsley says the cottager with the benefits of rights of common thinks that owning a cow or two will raise him above his brethren of the same rank, and give him confidence in a property he is unable to support, and then

> in sauntering after his cattle, he acquires a habit of indolence, a quarter, half and occasionally whole days are imperceptibly lost. Day labour becomes disgusting; the aversion increases by indulgence; and at length the sale of a half-fed calf or hog, furnishes the means of adding intemperance to idleness, another sale follows...

And the wretched man, now without his former subsistence, ends up on poor relief. Billingsley seems to have no knowledge of any industrious, sober commoners. He is adamant that his description is not exaggerated, giving the example of the village of Wedmore (where he was not a commissioner). Within the past twenty years, he says, more than 3,000 acres of rich moor land near Wedmore has been Enclosed, which when it was common land was unproductive due to its being wet for six or seven months of the year, and in the remainder of the year it was overstocked. This land is now worth 30s – 60s

---

* Auster tenancies were an unusual and ancient form of tenancy, mainly in the west of Somerset, based on whether a cottage had provision to bake bread.

per acre. He goes on to explain that Enclosures there were made by ditches, annually cleared out with the silt laid on the surface as excellent manure, and the employment in ditching and so on has meant the poor rate has been reduced, or at least not exceeded since the Enclosure. But 'it may be noted here as a fact, that in most of those parishes where no inclosure had taken place... the poor's levy has been doubled, trebled, nay quadrupled in the last 20 years'. He does not say which parishes without Enclosure he is referring to.

While the poor rate certainly rose in the period given it is difficult to be sure of all the reasons. On the one hand this was a time of inflation and extremely high prices for bread and other essentials which in itself will have caused distress; on the other, contemporaries such as Sir Frederick Eden produced evidence showing that pauperism was greatest in areas where Enclosure *had* taken place (though he did not mention Mendip or even Somerset).[20] And Arthur Young's *General Report on Enclosures* states that in the nineteen counties responding to his enquiries (again, not including Somerset) in every case the poor were injured.[21]

At the time of the Enclosures even relatively small tenant farmers could see the possible gain from Enclosure through having an apportionment for themselves. But from the social point of view Enclosure has a very bad press. Many of the most notorious examples were in areas like the Midlands with the taking of open fields already in cultivation, resulting in significant hardship for those who had been commoners. As previously noted, though, on Mendip Enclosure mainly affected waste land, poor quality and previously uncultivated, and, in this area the prime intention behind Enclosure does seem to have been for agricultural improvement.

In the folk-lore of the area Billingsley himself is said to have had a very negative reputation with the poor: it is claimed he was 'hated' for his part in the Enclosure of Mendip. Quite likely, but it is of course difficult to find written evidence for this, as most of the ordinary people themselves left no record of their situation, regardless of the level of their dispossession or their discontent. One of the few records we have comes from the popular poetry of the period, much of which was passed down orally and not written down until many years later. At least two versions of one such local poem exist, written in 1788 (incidentally the year before Billingsley first became a commissioner).[22] One version is said to have been composed by Jab Goul, in the other he is named as Nat Goul, a collier from Holcombe (a village just to the east of Ashwick).*
Both versions are in Somerset dialect. Goul gives a comprehensive account

---

* One imagines the two are one and the same – Nathaniel being his given name, 'Jab' an appropriate nickname.

of the dissatisfaction of the people on a range of issues including Enclosure, turnpikes and so on. The following snippet gives a flavour:

> They've tynd in all the common groun
> And pulled the cottage housen down.
> Wold men who once cou'd help a naiber
> Must now earn bread by thar hard labour…
> But wealthy power can't be withstood
> And girt men doan mind the public good.*

While many of the changes reported in the poem were not related to Enclosures – and all change seems to have been for the worse in Jab Goul's eyes – to say the Enclosures were unhelpful to those without land would be an understatement. On Mendip, though, the repercussions were far less for the common man than in most other parts of the country. Meanwhile, the very real agricultural improvements, implemented by Billingsley and others, were having an effect locally on production of grain and other commodities. Without them, bread would probably have become even more difficult to come by.

---

* Tyned = enclosed; Girt = great.

9
LAND ACQUISITION

BILLINGSLEY IS SAID to have acquired 3-4000 acres of land during his lifetime. Enclosure was an important factor in enabling much (but by no means all) of this. He had taken on the majority of his land by the 1790s, though tracing the early stages means returning to his young adulthood, while he was still working in the wool trade. In 1774 both his father John senior and his uncle Nicholas died. While there is no evidence that Billingsley inherited anything from his father, he did inherit some land from his uncle.* Prior to 1774 he does not appear to have held any land himself as main tenant, though he is named as a life on several tenancies and probably had the use of some family land.

Nicholas' estate included the tenancy of the Fosse House tenement – part of this was the Billingsley family home, Ashwick Grove. The story is told in greater detail in chapter 13 but, briefly, after the death of his uncle, Billingsley had to pay £380 to regain the lease on the property in his own name.[1] It amounted to two houses with outbuildings and just over 22 acres. In 1774 he was 27 years old: it was his first known sally into land and he did well in the negotiation. At that time he may not have been living at Ashwick Grove as he was described as 'a Clothier, of Shepton Mallet' (the location may have referred to his place of work, but usually refers to a person's home, and Ashwick Grove itself was in Ashwick parish). However, he had already started to be interested in farming and was beginning to experiment in methods of cultivation. It is quite possible that the land he was using was his uncle's, before he had acquired any in his own name.

Apart from the Fosse House tenement, Nicholas had also acquired other land in Ashwick over the years. By the time of his death in 1774 his land in

---

* As far as is known John senior did not hold any land during his lifetime, although he was mentioned as a life on several tenancies. It is likely, then, that any benefit passed to his son would have been in some other form, though Billingsley is not mentioned in his father's will and no evidence of any inheritance has been found.

that parish totalled at least 87 acres together with some cottages and other buildings, almost all of it leasehold. All this passed to his nephew when he died (with the income from some named fields reserved for charitable purposes, which Billingsley was tasked with administering). In addition to the Ashwick land Nicholas also left him a small amount of property outside the parish, so that Billingsley's whole land holding from his uncle's bequest was about 100 acres – still a relatively modest amount.*[2]

Up until this time Fortescue was very averse to selling the freehold of any of his Somerset land, preferring to keep outright ownership and benefit from the rents and other charges. But Billingsley was obviously in the market for outright ownership – he could see the advantages. On 17 December 1779, needing to write to Fortescue's agent on another matter, he took the opportunity to address it again:

> When I last had the pleasure of Interview with you at Castle Hill I desired to know if his Lordship would sell the Fee of any of his Estates at Ashwick as I was desirous of purchasing what I hold under his Lordship --- You could not say whether he would or not but promised to write me on the subject --- As I apprehend the Conversation has escaped your memory on that Subject in your reply to this Letter and you will oblige.[3]

This appeal was still unsuccessful. He had to wait another twelve years for Fortescue to decide to sell outright.

Meanwhile, as Billingsley was moving out of the wool trade, he began to acquire other land. In March 1776 he entered into partnership in the Oakhill Brewery with James Jordan. His first purchases were for this partnership, rather than for himself as a private individual. It seems most likely the land purchases described below represent, at least in part, his investment in the business. Ubley, on the Mendip plateau, is about twelve miles from Oakhill, so quite a distance to travel frequently on horseback or by carriage. The choice of location was probably due to availability in an area where Jordan had family links.

One of the first Mendip Enclosure acts was for the parish of Ubley (1773).[4] As often happened some of the allocated common land became available for purchase soon afterwards. Early in 1776 Jordan and Billingsley entered into negotiations for several plots (with Samuel James 'of Okle' as trustee).†[5] The purchase happened gradually and with many separate legal

---

* The Fosse House tenement also included land in Shepton and Stoke Lane parishes.
† James was related to Billingsley's mother.

processes. Most of these plots were small (less than ten acres), being individual allocations to minor land owners, which many of the recipients then needed to sell to avoid the expense of Enclosure. To complicate matters some plots were jointly owned by several people. As just one example: a field of five acres had been allotted to Philip Dirrick, now deceased, who left it in equal parts to his wife and seven children, meaning negotiation and legalities with all eight in order to transfer ownership.* Adding all these small plots together, Billingsley and Jordan bought just over 142 acres in 1776, then further land in 1777, leading – by rough and uncertain estimate – to in excess of 190 acres, for a total cost of £2,371 11s 8d – all paid by Billingsley, and all purchased freehold. The various plots were 'contiguous the one to the other'. The partners immediately used the land as security to raise three separate mortgages totalling £1,200 in 1776, rising to £3,000 in 1777, moving from one mortgagor to another with some frequency. No doubt most of the land was swiftly rented out for an income, although some must have been retained by Billingsley himself, who carried out his earliest documented trial there, resulting in his first Bath and West premium (for growing carrots, see chapter 6).

It must have seemed to Billingsley that land was a good investment. In 1783 he was able to also acquire an entire manor in the immediate area, this time for personal ownership and not with Jordan. For the purchase of Hazel Manor Billingsley went into partnership with his nephew, Thomas Parsons, son of his sister Mary. The Parsons family lived in Publow, but on the Hazel documents Thomas, aged 26, is recorded as 'of Ubley' so had evidently moved there by 1783, possibly onto the land his uncle had bought with Jordan (see below).⁶

The manor of Hazel is spread over two parishes: Ubley and Compton Martin. It is a very old manor but with few historical references. In 1760 it was sold at auction to Thomas Mitchell for £1,845, and later passed down to his son, also Thomas. The younger Thomas Mitchell sold the manor to Billingsley and Parsons: the sale of the 'Lordship and Manor of Hasill alias Hasely' being concluded on 6 November 1783. There is no mention of a dwelling. The indentures marking the sale show Billingsley as having a 'moiety' share and Parsons the other moiety. According to the document the lord of the manor was required to hold a regular court leet and court baron, be responsible for 'View[ing] the Frankpledge', and holding the 'Goods and Chattles of Felons, Fugitives, Outlaws and Attainted Men'.† Ownership of the mines, 'open or

---

* Philip's sons included the interestingly named 'Shadrach, Meshach and Abednego' (as in the biblical story).

† The Frankpledge was a medieval system in which a group of ten households were jointly

undiscovered' was also included in the purchase – this was potentially very lucrative and something Billingsley subsequently made use of. The money was raised by the partners at the time of the purchase, through a mortgage for £3,000 at 5%. They each paid Mitchell £890, making a total of £1,780 (so £65 less than Mitchell senior had paid for it in 1760). This much is clear, however tracing the full details of the whole acquisition is extremely tricky, the result given here admittedly tentative and incomplete.

The Hazel estate was later said to extend to '1000 acres of Ground or more'. It was the largest block of land he would ever take on, a major acquisition by any standard. However, this figure must include land other than the manor itself.[7] While the exact acreage of the manor is unknown, either it was very much cheaper per acre than the Ubley land, or a good deal of extra land must have been charged for separately from the original manor. One example of this extra land is the charmingly named and relatively small plot known as 'Cookow Thorn'. The partners also acquired what seem to be two working farms in the vicinity, comprising 'two messuages, two cottages, two barns, two stables, two gardens, two orchards' plus land totalling 256 acres and 'all Manner of Cattle'.* Land in the parishes of Ubley, Compton Martin and Chewton Mendip, some of it still unenclosed in 1783, all became part of what was later regarded as the Hazel estate. The various parts (plus the Ubley land purchased in 1776-7) were situated close together, leading to a coherent estate which could be managed as one.[8]

Although it is believed there was once an Elizabethan manor house at Hazel this building must have disappeared by the time of the Day and Master's 1782 map, which shows no house in the vicinity.[9] However, something seems to have been built there by 1783: according to Atthill this is very probably the house initially called Ubley Hill Farm and subsequently known as Hazel Manor.† Presumably (though this is conjecture) it was built by the partners. As Thomas Parsons is recorded as 'of Ubley' perhaps he lived there.

Tracing the ownership down the years reveals that in 1787 Parsons pulled out of the partnership of the Hazel estate, conveying his moiety to

responsible for each others' conduct and behaviour.
* These farms must have been relatively new as prior to the Enclosure act there appear to have been no buildings on the land.
† Long past Billingsley's day, Hazel Manor house, which had been much altered in the interim, suffered a major fire in 1929 and was never rebuilt. A photograph shows that its gothic window and door surrounds had been removed at some time before 1924 and relocated to Fernhill Farm – another Hazel estate farm probably built by Billingsley. Incidentally, the names of the farms in the area seem to have changed and been reused over time, the name Ubley Hill for example now belonging to another farm.

Billingsley. Recorded in the same document is another rearrangement of the mortgage with a different mortgagor, now amounting to £2,500 at 4.5%. In 1803 Jordan likewise relinquished his half share in the Ubley lands.[10] As he retired from the brewery about this time it was presumably a way of taking some of his share out of the partnership (perhaps he used the proceeds as a retirement fund). It also meant that Billingsley now had a larger share in the brewery.

In 1795 he consolidated the mortgages for the two sections of land (Ubley and Hazel), passing them to Thomas Jolliffe of Kilmersdon. By 1803 Billingsley had mortgages totalling £8,000 on land in the area.[11] When he died in 1811 he still held all these lands – apart from that in the brewery trust which he had withdrawn from in 1810, repaying £1000 of the mortgage at that time. The rest of the land was still mortgaged to Jolliffe, with a total debt of £7,000 in 1811. Billingsley – and his solicitor, Edmund Brodrib – must have had a difficult time over the years, keeping track of all the changes to the various purchases and associated mortgages.* In 1818 (after his death) Billingsley's trustees advertised the Hazel estate for sale: 'upwards of 1007 acres, statute measure, and about 70 acres of woodland…abounding in lead ore and lapis calamaris [calamine]'.†[12] The estate was sold to Charles Wilkins Esq – who promptly mortgaged it with Jolliffe again.[13]

By the early 1780s Billingsley's method for land speculation was well-established. His system is obvious: initially using his funds from the wool trade (and perhaps some from his uncle), he bought land, raised a mortgage on it, used the proceeds to buy more land, and so on. Meanwhile, he rented out most of the land – presumably helping to pay the mortgage – in some cases retaining a portion for his own use. Simple but effective.

Billingsley's final purchase in 1783 was of the manor of Stoke Lane, which extended to 480 acres – meaning that in this year he bought more land than in any other in his life. While, probably to his frustration, he still did not hold his Ashwick land outright he did at least have a long tenancy on it. So, when land adjoining Ashwick Grove came up for sale, he purchased it 'In Fee', stretching his home estate eastwards in the process. He bought this land from Thomas Horner of Mells; there is no record of the price.[14] No sign of a mortgage has been found, although, in view of the amount he was buying in that year he probably did raise one.‡ A good deal of the newly purchased

---

* Brodrip's name is spelt this way on the documents, but as Broderib or Broderip elsewhere.
† It is not clear whether the 70 acres of woodland are in addition to the 1,007 or part of it, but it is assumed here that it was additional.
‡ Also in 1783, Billingsley sold the logmill in Stoke Lane that he had purchased with his

land was either already subject to leasing agreements or was soon rented out, though he again retained some for his own use, as pasture and water meadow (see chapter 12).

Billingsley soon began to acquire yet more land through various Enclosure acts. For example, as already mentioned (in chapter 8) he was awarded at least twenty acres from the Shepton Mallet act of 1785 (for which he was not a commissioner) as a result of his existing holdings of several small fields in the manor.[15] He also held the unusually small (and obscure) manor of Elverstoke, somewhere to the south-west of Oakhill. The exact position and size of most of his plots is unknown, but by a conservative estimate his final holding in Shepton parish from all sources (including Nicholas' bequest) must have totalled in excess of 80 acres (as well as some brewery trust land lying in Shepton parish, the extent of which is also uncertain, but is not counted in his personal holdings, see chapter 8 for more detail).

Meanwhile, Billingsley was still waiting to buy the freehold of his home at Ashwick Grove. In 1791 the moment finally arrived. A major survey of Fortescue's land in Ashwick was undertaken, with a view to a sale. Billingsley was one of the main purchasers, buying up most of the land and property he had held leasehold from Fortescue, plus some more.[16] A copy of the survey is endorsed with full details of the purchasers and prices. The properties Billingsley bought included – at last – his home at the Fosse House tenement, plus additional land in the vicinity: a combined total of 14 houses and 77 acres.* For all these he paid a total of £275 8s (see the note below and chapter 13 for more detail). Having had long leases for most of this (being effectively a sitting tenant) he paid much less than he would have done for the leasehold tenancies, or for unencumbered land elsewhere. In 1803 he bought further Ashwick land at Fortescue's auction of Ashwick manor (but not all the land he had previously rented), meaning he then owned about 115 acres within the parish.†[17]

During these years land continued to become available through other Enclosure acts. Billingsley is said to have 'reserved' plots in some, but how

---

friend John Bowles in 1773; it is probably a coincidence that this mill was within the manor.

* The Fosse House Tenement land included: item no 1 being two houses with 27½ acres, for £46 10s; item no 2 is Ashwick Grove house. In addition to this he bought another 12 houses/cottages and a further 49½ acres locally for the cost given.
† At the time of Billingsley's death what were known as his 'Ashwick Farms' contained about 168 acres, but this amount included land in Shepton Mallet and Stoke Lane already accounted for above.

much and exactly where is again uncertain. In 1794 he bought the 124 acres at Green Ore, costing him £1,115 13s 6d (see chapter 8).[18] From the Chewton Mendip Enclosure he acquired a further 80 acres. Over the years Billingsley had also gained land from other Mendip Enclosures, some allocated, some bought and some exchanged, each in small amounts as a result of the Hazel manor land he already held. These must have added at least another 100 acres. Eventually, through all these different means, he owned land on the Mendip plateau in the parishes of Ubley, Compton Martin, East and West Harptree and Chewton Mendip. He does not, though, seem to have bought any land on the Levels following the Enclosures there, preferring to keep his interests closer to home.

It is clear from the above that 1783 was the peak year of Billingsley's land purchases, thereafter he bought only relatively small amounts and for specific purposes. The Ashwick land, for example, including that of his home and nearby ground, was obviously for his own private use, being known as his 'Ashwick Farms'; the land at Green Ore he converted to a typical Enclosure farm, using it to demonstrate his ideas and methods in the essay he wrote for the RSA (described in chapter 8). What is not obvious is why he stopped investing in large amounts of land after the mid-1780s. He had been through a period of seeming to consider land to be the best investment. So, did he stop investing in it due to the difficult economic situation and increasing land prices, or was his outlay on other things increasing but not his income? Perhaps, if he still wished to speculate and had sufficient money available, he now believed that other methods, such as investment in canals or mining, would be preferable. Or, he felt he simply had enough land to cope with and was less interested. The renovation of Ashwick Grove starting in 1791 must also have been an expensive business, perhaps limiting the amount he had to spend on other things (see chapter 13).

Determining the amount of land Billingsley acquired during his lifetime is problematic and ultimately unsatisfactory. We have only partial information. Although land and property transactions are among the most common records for this period, not all such records have been saved. Those that have been discovered so far do not always record the extent of the land conveyed in acres, and even where field names are given they have often changed or disappeared. In some relevant cases there is very detailed evidence of the position and quantity of the land; in others, large areas are frustratingly vague in extent, or parts apparently overlooked. Some purchase records may or may not be part of, say, the larger investment in Hazel Manor. In attempting to provide a total the figures given here count his definite acquisitions, but they may not be the whole story. At least it is fairly certain that any mistake in the result is an undercounting, rather than overcounting.

The extent of Billingsley's personal acreage, that we definitely know about, adds up to some **2,056** acres in total.* It is an impressive amount for a man who started with an inheritance of under 100 acres, almost all leasehold. Yet it has been said – repeatedly – that he accumulated 'some 3-4000 acres'. So wherein lies the discrepancy? Either a large amount of land has been missed – quite possible, despite a diligent search, but rather more than expected from just, say, further Enclosure reservations. Or is there another explanation? The most likely answer is that people regarded land that he was cultivating for others, particularly Earl Waldegrave, as if it were his own (see chapter 16). He certainly undertook an extraordinary amount of work in reclaiming land which was not his own. Similarly, the brewery trust land and small amounts of land in trust for the presbyterian chapel may possibly have been counted as his own. If both these areas of trust land are included in his holdings then they would add at least another 206 acres.

William Matthews, in his capacity as editor for volume X of the Bath and West's *Letters and Papers* (1805), claimed that Billingsley 'enclosed and rendered productive between 3,000 and 4,000 acres on his own account'.[19] The 'on his own account' is confusing. It is generally thought to mean that he owned the land, though it may have been meant 'by his own actions'. If the 1,500 acres he 'enclosed and rendered productive' for Waldegrave and the 206 acres of trust land are added to the 2,056 acres accounted for above, then the total would become **3,762 acres**, well within the amount Matthews and have others claimed. So – that he personally *enclosed and cultivated* between 3,000 and 4,000 acres is altogether plausible, but it seems unlikely that he personally *owned* quite that.

---

* Billingsley's confirmed acquisitions total: the Hazel estate 1077a; Manor of Stoke Lane 480a; Shepton Mallet 80a (including from Enclosure); Ashwick land 115a; Green Ore 124a; Chewton Mendip Enclosure 80a; various other Mendip Enclosures 100a; making a total of 2,056 acres. This does not include land in trust, such as that for the brewery and land at Masbury (ca 3 acres) and at the Wells (St Cuthbert Out) Enclosure (ca 13 acres) held for the nonconformist chapel: combined total 206 acres.

10

NAVIGABLE CANALS

AFTER 1783 BILLINGSLEY's speculative interest gradually lessened, moved away from land purchase and after a short pause was reignited by canals and to a lesser extent mining. Following the success of England's first true canal, the Bridgewater Canal (authorised by Parliament in 1759), 'canal mania' gripped England. The period 1770-1830 is widely regarded as the 'golden age of canals', after which the railways arrived. Building a canal required an act of parliament: by 1774 there had been 33 Acts, mainly in coal-mining areas in the north and midlands, places where both commerce and topography were better suited to canal building than in the south-west. In order to compete nationally the south-west region needed its own canal system, but action only began in the 1790s. Flurries of proposed canals were then announced, with frequent calls to establish committees or encourage shareholders, Billingsley an enthusiastic participant in the trend.

As in the case of road transport there was no government financial support. Once a canal had its own act of parliament the company could sell shares and buy land. The huge costs involved prompted a special method of raising capital, a variant of joint stock companies, with shareholders (described as 'Proprietors of the Company'), and a system of 'calls'. Investors made a down payment, then pledged further payments to be paid in when required: funds were called on as needed. This limited investors' exposure in case the project was abandoned unfinished; a wise approach, as many canal proposals foundered through lack of finance, though large numbers did still go ahead. However, the south-west was a little late to the table: by the 1790s the prevailing economic situation meant that investment on such a scale was becoming increasingly hard to come by, and even some canals that were finished still never paid a dividend. For Billingsley himself canals were probably the least successful of all his investments. In some cases he definitely lost money.

In the south-west other challenges were at least as pressing as raising the finance. In north Somerset particularly the topography is very unsuited to canal building; there was proportionally less heavy industry needing water transport; landowners were frequently obstructive, refusing permission for purchase of their land; in many areas sufficient water was not as easily available as further north; and technological issues abounded. Technological successes could be wonderful, but the difficulties inherent in canal building meant failures were rather too frequent, or the expense so great that it spelt the end of the project. Against this backdrop, for the years between 1790 and 1810 – the last two decades of Billingsley's life – there was still profit to be made. During these years enthusiasm was such that a bewildering number of canals were proposed within Somerset alone. Only a few were ever built. As a speculator by nature, and interested in both trade and water projects, Billingsley naturally became involved in the committees setting up projects. He is not known to have contributed in any practical way to the engineering side of the enterprises (though water management definitely attracted him), confining his involvement to the committees and investment. It is worth noting, though, that while 'pure' speculation was common in the early days (for example, making a profit by acquiring shares early then selling them on before building had even started), wherever he was chairman Billingsley took steps to ensure it could not occur, whether through principle or consciousness of potential loss of profit.

Billingsley was active in at least four canal projects, which had varying degrees of success:

The Kennet and Avon Canal
The Somerset Coal Canal
The Dorset and Somerset Canal
The Bristol and Taunton Canal.

He may also have invested, but not been active, in other projects, though no details of any can be verified.* Those canals that he was involved in were all local to his area, as was his usual practice. Taking each in turn his part in it is detailed below, but the often very complicated stories – committee squabbles, landowner obstruction, engineering challenges, changes in route

---

* For instance, a John Billingsley invested £200 in the Stroudwater Canal in 1776. As wool was then declining in Shepton Billingsley may have wanted to speculate outside the trade. Also, his family still had strong links with Awre, just across the river from the end of the canal. But it was a long way away from his home ground, and would have been his first, very early, canal investment. It has not proved possible to confirm that this was John Billingsley of Ashwick Grove: there were others in the area with the same name, so it may well have been another man.

and difficulties in construction before the canal was even operational – are all kept minimal.

The Kennet and Avon Canal was initially proposed under the name of the Western Canal (the change of name was to avoid confusion with the Grand Western Canal).[1] While it is said that Billingsley was 'actively engaged in the promotion of the Kennet & Avon Canal' it is not clear to what extent.[2] This project may have specially interested him as it would link up with the proposed Somerset Coal Canal, opening new and distant markets. Billingsley may possibly have been present at one or more of the early meetings, but he was not a member of the committee.

The Kennet and Avon was intended to close the gap between the River Avon, going from Bristol to Bath, and the River Kennet which linked Reading to Newbury and then on to the Thames. In view of the amount of trade between Bristol and London, a link there was plain economic sense and had been repeatedly suggested over many years. At the time the project began it was likely to be easily profitable. But plans for the link were long delayed. At the first meeting in April 1788, the Hon Charles Dundas (1771-1810, MP) was elected chairman. John Rennie was approached as engineer; his suggested route would cost £213,940.* In November 1790 'Subscription Books' were set up, with the intention of collecting £75,000. It is believed that Billingsley invested at this early stage, and that he kept his investment for the rest of his life, but it is not known how much he put in. Initially there was opposition to this canal, which was eventually overcome by 'either patient negotiation or bribery'.[3] The delay had caused frustration among investors and businessmen resulting in an attempted coup, which forced Dundas to move forward. In 1793 the project finally restarted, but with an altered route. By this time the Somerset Coal Canal had also chosen Rennie as engineer and was proceeding to an act. That canal's contribution of freight would increase their profit, in turn increasing optimism for the Kennet and Avon.

In the event both canal projects were granted approval by acts of parliament on the same day, 17 Apr 1794.[4] Despite difficulties with construction part of the Kennet and Avon was open and in use from 1801, although near Bath freight still had to be unloaded at the bottom of a steep section and transported uphill by railway until locks were complete. Engineering challenges meant it was the end of 1810 before the whole length of canal was open. It had been sixteen years in construction. Once open the canal was financially successful, in large part due to coal traffic from the Somerset Coal Canal.

---

* John Rennie FRS (1761-1821), Scottish civil engineer who specialised initially in canals; later in bridges, docks, harbours and lighthouses; a pioneer of structural cast-iron.

Unfortunately for Billingsley, though, the profits began only in the year before his death. The canal remained profitable until the time of the railways in the 1840s, taking tolls of upwards of £50,000 in some years.

Meanwhile, the Somerset Coal Canal was being promoted by the mine owners of the north Somerset coalfield as a cheaper means of transporting coal to their main markets in Bath and Wiltshire. Until the 1790s Somerset coal could still only be transported by pack-horse or cart, which seriously limited the quantities that could be moved, keeping the price high. The movement of coal damaged the roads: as early as 1617 the inhabitants of Stoke St Michael were complaining that 'by reason of many coalmines...the highways are much in decay and grown very founderous'. By the eighteenth century more coal seams were discovered at Radstock, together with the likelihood of cheaper coal soon coming from south Wales, which would increase the traffic. In 1792 an act for the Monmouthshire Coal Canal was applied for, which if passed was sure to bring coal to the area at a lower price, further concentrating minds. Billingsley took a much more personal interest in the coal canal presumably because it was local, plus he was involved with Earl Waldegrave's mines at Radstock (see chapter 16).

The north Somerset mine owners called the first meeting in December 1792.* Proceedings were reported in the *Bath Chronicle*:

> On Monday se'ennight a meeting was held at the Old Down [Inn], Mr Billingsley in the chair, to consider of the propriety of making a navigable Canal from the several collieries to this city, when a Committee was appointed to wait on the proprietors of land through which it is to pass for their approbation.[5]

The next meeting was in February, 1793, again at Old Down, but with Henry Hippisley-Coxe Esq in the chair.† The meeting agreed that a navigable canal 'will be highly useful and beneficial to the public at large...as coals will be rendered much cheaper', especially in Bath.[6] A committee formed of twenty 'Proprietors of the Company' was set up, including Billingsley and the colourfully named Eleazer Pickwick of Bath, whose interest in the project

---

* It is likely that Billingsley had already invested in coal by this date, but no record can be found. His only verified investment did not take place until 1797 (see chapter 11). Yet all the proprietors of this canal – of which he was one – were said to have 'interests in coal', so it is puzzling not to have found him involved in any mine prior to 1792.

† Henry Hippisley-Coxe of Ston Easton Park (1748-95), MP and Sheriff for Somerset; a local land and colliery owner. James Stephens soon became a more permanent chairman.

proved vital in the later years of the canal's construction.* The committee was empowered to employ engineers to survey, plan and estimate expenses for branches to the northern collieries, and to look at the possibility of a branch to Radstock, to serve the southern collieries. The initial intention was to connect the northern collieries with the Kennet and Avon at a point near Monkton Combe. The engineer chosen was John Rennie who was already working on other canals (including the Kennet and Avon), so had both expertise and credibility in this area.

Rennie's route involved both branches. For the surveys he was assisted by William Jessop and William Smith (the geologist), the latter going on to be involved in the construction itself.† Starting from the Kennet and Avon end the route chosen had a 'main line' as far as Midford, where it split into the northern branch to Paulton and Timsbury via Dunkerton, and the southern branch to Radstock. Each route required a rise in level of 135ft after Midford; a three-quarter mile tunnel was also needed on the northern branch. Rennie's estimated cost of £80,000 was accepted unanimously. A subscription was entered into, with 300 shares for £100 each reserved for land-owners in certain parishes whose land the canal would pass through – a good way of ensuring landowner support. Any shares outstanding would be offered to existing subscribers or 'disposed of amongst such persons of responsibility as may then be present'. Parishes named in the list included all those in the northern branch, but not those in the southern branch, so presumably plans for the southern were not as advanced at that stage.

Writing in 1794 Billingsley seems relatively optimistic of the northern branch, but considered the southern branch to be less definite.[7] In April 1794 the (as it was then known) Somersetshire to Bradford Canal Act was passed, as already noted on the same day as the act for the Kennet and Avon canal, a project much longer in the planning.[8] The act listed the 182 initial shareholders

---

* Eleazer Pickwick was a wealthy vintner from Bath, a relative of Moses Pickwick who owned a stage coach business there. The family – and stage coach – are immortalized in Charles Dickens' 'The Pickwick Papers'. Other members included Billingsley's friend Richard Perkins and Jacob Mogg.

† William 'Strata' Smith (1769-1839), geologist. Working in the Somerset coal field and canal he recognised that layers of rock follow a predictable pattern in strata relative to one another, and that layers could be identified by the fossils they contain. Credited with creating the first nationwide geological map of any country. In 1799 Smith's contract was discontinued (probably not for the reasons publicly given). William Jessop (1745-1814), engineer, initially worked with Smeaton then came south and worked on canals and docks. Other engineers who worked on the project included John Sutcliffe and Robert Whitworth.

('Proprietors of the Company'), including Billingsley, who invested £500. More investors joined later.

Work started on the northern branch of 10.6 miles.[9] The route eventually chosen meant the canal joined the Kennet and Avon at the Dundas Aqueduct. To save money Rennie altered the line of the northern branch so as to avoid the tunnel, but dealing with the fall of 135ft in an area of insufficient water proved to be a major engineering challenge. For some reason the committee did not want to use conventional locks and looked for another solution. Robert Weldon's innovative 'Patent Hydrostatick' or caisson-lock was put forward to cope with both the fall and the lack of water in the problem area.[10]

In engineering terms Weldon's lock was the space craft of its day – a watertight chamber in which a loaded barge would descend 46ft to a lower level, with the bargees inside. Successful public trials – one watched by the Prince of Wales – proved that the system worked in practice, but problems and accidents finally led to it being abandoned (it was eventually discovered that its failure was due to unstable soil rather than any deficiency in the design itself). Billingsley must have been an interested observer of this invention – quite probably present at one of the trials. He included a description of it, written by Weldon himself, together with a diagram, at the back of his *General View..*, one of only two contemporary accounts known to have survived.[11] Within a short time two further solutions to the sudden drop in level also proved impractical, the second causing serious delays. Rising costs led to another act in 1796, to enable further financing and change of route.[12]

By June 1798 'a proportion of not less than two thirds of the canal is now in a state of perfect completion'. However, goods still needed to be offloaded then reloaded enroute. By November 1801 the northern section was complete but the raising system being currently used needed to be changed to a flight of locks. Although the canal had already successfully reduced prices of coal in Bath the canal's own finances were now critical. At this stage the bankers refused to lend any more to the project, while some subscribers were in default with payments and under threat of being sued for the money. Eleazer Pickwick saved the day, lending a crucial £10,000. In order to finance the necessary flight of locks the organisers intended to apply a toll of 1s per ton on all traffic – vigorously opposed by neighbouring canals. So a third act was sought, passed on 30 April 1802, enabling the organisers to raise a further £45,000 (known as 'the Lock Fund').[13] This was achieved by the Wiltshire and Berkshire canal, the Kennet and Avon canal and the Somerset Coal canal each putting in £15,000; extra tolls would be paid to repay the fund. Then again in 1803, by which time Pickwick had been elected treasurer, there was another financial crisis. Once more Pickwick stepped into the breach and produced another £11,000 so work was able to proceed. Tramroads were used to complete the route to Bath, until a flight of 22 locks was finally built in 1805. By November 1807 the canal had the enormous debt of £50,514 (counting in Pickwick's loan of £21,000). The committee, presumably still including Billingsley, was having a difficult time. However, once the locks were in place things improved immeasurably so that by 1810 the lock fund was issuing dividends.

The southern branch, planned to start at Radstock basin also suffered from lack of water, plus some landowners were obstructive. Although an initial section of canal was built this branch was never connected to the main line by locks, a tramroad taking goods onwards was installed, meaning unloading and reloading. This branch had lost out on expenditure to the potentially more lucrative northern branch from the outset and was never very successful. In 1814 the whole of the southern branch was replaced by a tramway. Meanwhile the northern line prospered. Billingsley seems to have retained his investment throughout the canal's vicissitudes, but it is not known whether he remained on the committee for the duration. By the 1820s the remaining part of the coal canal carried over 100,000 tons of coal annually, making it then one of the most profitable canals in the country – but after Billingsley's death.

The Dorset and Somerset Canal was also initiated in 1792.[14] This project was a disaster. What should have been a great asset to the two counties' otherwise poor transport links soon ran into construction and financial difficulties and was abandoned. Unfortunately for Billingsley he took an early shareholding, eventually investing a total of £3,500. This canal would have been as relevant to him personally as the Somerset Coal Canal – not just for his investment and as a member of the committee, but also because the Nettlebridge branch would have gone within a mile of Ashwick Grove. It could, for example, have been used to transport the Oakhill Brewery's beer, which would have been a great help not just to the brewery but also to the local roads. The debacle probably represented his biggest loss from canal speculation, it may well have been the biggest single financial loss in all his speculation. It was also a loss to the prospects of his own area – and a loss of his time and effort in what was the busiest period of his life.

The idea behind the Dorset and Somerset was to connect Bristol to the English Channel, thereby avoiding the dangerous passage for goods along the south coast and round Land's End, connecting inland areas of these counties to the rest of the canal network.[15] It was one of several projects with the same aim put forward. The major trade would be coal coming from the north and clay and wool from the south, plus movement of other heavy goods such as iron and grain. The main line was to go from Bath, meet the Kennet and Avon, then via Wincanton to Poole on the south coast. A branch line from Frome would go up the Mells valley to Nettlebridge, thereby connecting with that part of the Somerset coalfield that the Somerset Coal Canal did not reach. Whether Billingsley was instrumental in ensuring that branch was included we do not know. He may have been a prime mover in the whole project; in any case he was very active in the committees.

The first meeting was at Wincanton on 10th January 1793. It is unlikely that Billingsley was present at that meeting and he was not elected to the initial committee. It was immediately obvious that the route would be contentious. Further meetings in different places followed quickly, each meeting arranged by a faction promoting a slightly different route. The main decision came down to whether the canal should pass through Wareham or Blandford or both. A new committee for the project was established on 7th February 1794, again at Wincanton, tasked with deciding the route. This committee was described as 'a masterly compromise' as the various factions were all represented. Billingsley was elected a member: he was seen as useful in already being connected to both the Kennet and Avon and the Somerset Coal Canals, as well as having other local interests such as the brewery.

The route decided included both main lines: the canal divided south of Wincanton, with a side branch to Hamworthy, then rejoined south of Wareham and on to Poole. There were still several changes to come. Meanwhile, in 1795 it was decided that 'in order to avoid improper speculation... all future subscriptions should be confined to the proprietors in the first instance'. At that point things must still have seemed positive for the existing investors to make good profits; as it turned out they were the ones to lose money rather than those they excluded. With projected costs increasing; subscribers were asked for a 3% deposit. The canal would no longer run from Bath, but would meet the Kennet and Avon further south-east, costs of these changes not mentioned. Costings for the branch to Nettlebridge do not seem to have been a major concern at that time.

The act was passed in March 1796, by which stage costs were estimated at £200,000.[16] The route specified ended several miles short of the coast at Poole. It is said that a major landowner, Lord Rivers, was adamant that the canal should go no further south than Shillingstone (north of Blandford Forum), otherwise he would withdraw his consent.* In those days important landowners were all-powerful, so this effectively prevented progress. Although

---

* Lord Rivers was Lord of the Manor of Shillingstone, so presumably it would have affected his land, or his view – or he foresaw profits from unloading goods in his area.

on the river Stour, Shillingstone was a good 20 miles from the open sea. The act had authorized the raising of £150,000 with the power to raise a further £75,000. Although this was more than the estimate at the time of what was needed, yet it would prove to be totally inadequate. An undated but near to contemporary report summarised the canal's financial situation thus:

> the spirit of speculation was so completely changed by the altered condition of the country, the influence of the French Revolution, the large sums raised by Government on loan and the high prices which accompanied those events, that not more than £70,000 was subscribed, and out of that sum, only about £58,000 was actually received by the company.[17]

The act also stipulated that the branch line from Frome to Nettlebridge should be completed first. In view of the geography of the valley this decision was a mistake. It was evidently intended to take advantage of lucrative profits from the coal trade which would then fund further building. But this branch was technically the most difficult – and therefore the most costly – part to build. Had the route been started from the easier to build sea end it might have had more immediate success.

Work began on the Nettlebridge branch in several disassociated stretches of about two miles each. Financial difficulties began early. By 1798, given the technical difficulties of this branch and with only a fraction of the necessary subscriptions received, it is not surprising that the project was already overspending. Further money needed to be raised: a call of £10 per share was made.[18] An early list of subscribers shows 230 names, but in April 1800 more shares were being sold. Work on the Nettlebridge branch progressed, so within a year three further calls were made.

The Nettlebridge branch line was beset by the usual Mendip sudden changes of level. For example, the level of the route near Mells falls by 210 ft in less than a mile, so some creative solution was obviously required. Fussell put forward his 'Patent Balance Lock', previously offered to the Somerset Coal Canal. A model established that it worked in principle. It was trialled very successfully on several occasions, before a large crowd and to great acclaim. In October 1800 the *Bath Chronicle* reported that:

> a great crowd of spectators: among them were many men of science…after repeated operations without the least difficulty or mischance… were unanimous in declaring it to be the simplest and best of all methods yet discovered for conveying boats from the different levels.[19]

One imagines Billingsley among the crowd. Although in trials the lock had lifted boats by only as much as 20 feet it was said to be capable of 40 ft. By January 1801 the balance lock was 'in an actual state of work' on the canal. Five locks were planned in series to meet the required 210 ft. To give an idea of the building that had occurred along the valley, Robin Atthill quotes a list (from a pamphlet of 1825) of almost 100 completed features, including a 'noble and stupendous aqueduct' – all within eleven miles.*[20] But it was still nowhere near finished. All along the valley today there are puzzling features of periodic building or excavation, ending just short of Nettlebridge. The canal's whole course can still be traced, though the various sections do not join up.

All the money raised – plus an extra £10,000 – was spent on the branch line, which was still some way from being finished when work stopped. No work had been done on the main part of the canal. By April 1802 not only had the money run out, a debt of £10,00 had also been accrued, so in March 1803 a special meeting of subscribers was called to try to find some way to continue. Billingsley's friend Richard Perkins was among those who had called for the meeting, but Billingsley himself was not (presumably he realised further expenditure would be fruitless).[21] Momentum was lost and work abandoned. In 1825, after Billingsley's death, renewed efforts to revive the project also failed.

The Bristol and Taunton canal was yet another doomed proposal, one among several put forward at the time. Details are few, but serious proceedings had begun with 'a very numerous and respectable meeting', also in December 1792, this one at the Swan Inn, Wells, with Billingsley once more taking the chair.[22] The proposed canal was not initially given a name, simply described as 'a canal from Bristol to Taunton'. A committee of 21 named members was set up – including Billingsley himself. The route was not described in detail, except to say that it would run from Bristol and join the Grand Western canal at Taunton – the committee was directed to confer with the latter as to the place (this was one of Billingsley's tasks). Subscriptions were immediately entered into, with clear financial rules laid down from the start. It is very probable that Billingsley subscribed; if so, we do not know how much.

Despite the confident start things did not go smoothly. It seems that a group in Taunton had already planned something similar and had surveyed a route themselves, involving them in expense. This may not or may not have been known to those behind the Wells meeting. The Taunton group swiftly published a notice claiming that the meeting at Wells had been called

---

* The aqueduct is still known locally as 'the Huckyduck'.

with insufficient notice and a 'studious concealment of its object'. Those from Taunton who had attended at Wells 'had not been candidly dealt with'. Hence, they felt that nothing should stop them from going ahead with their own plans, which would be better for those in Taunton. It is probable that this initial disagreement was behind the slow progress of the venture. Some discussion between the two groups must have taken place, as another meeting was called, at which plans and estimates as well as surveyors' reports would be laid before them and they would then take the opinion of the subscribers as to whether to proceed.

Something must have been agreed, as the Bristol to Taunton canal committee that had been appointed at Wells in December 1792 was then reappointed as of June 1794 (this was apparently the same group that had assembled at Wells under a newly agreed name). In the October another general meeting proposed a conference to be held at Bridgwater in November of those concerned with three Somerset canals: the Bristol canal, the Bristol and Taunton canal and the Chard canal.[23] The multiplicity of names is confusing, but indicates the state of canal mania at the time. At this stage there is no further mention of Billingsley who seems to have withdrawn; perhaps he had become exasperated by the complications and lack of progress. It is not known what happened at the joint meeting, but again there was no progress for some time. In July 1795 a notice announced another general meeting in Bristol for the Bristol and Taunton canal. In December the subscription book was open and potential investors were advised to sign up by December 1795 or be excluded.

On 23 February 1796 the report on the parliamentary bill of the Bristol and Taunton Canal was received by the Commons. In the debate on this bill, Sir William Young argued that 'the measure would prove a private inconvenience and be of no publick advantage'. He then moved that the debate be adjourned for three months; the motion was passed.[24] This canal faded into oblivion. No doubt investors lost at least a portion of their money, if not all. In 1810 another Bristol and Taunton canal was initiated, but seems to have been a different project.[25] There is no evidence that Billingsley was either involved in or invested in this new venture, which began – as it turned out – only shortly before his death.

Several reasons combined to affect the lack of success of the Somerset canals in which Billingsley was most involved. The main factors were the downturn in economics, the difficulties of terrain and the behaviour of some of the other individuals concerned. Whichever was the most important, the net effect was that this was the least productive of his various interests.

11
Mining

Due to its potential for lucrative returns one would expect Billingsley to have been very interested in mining, especially as his own area was commercially active in it at the time. But as far as is known it was not until the 1790s that he became involved to any practical or financial extent, and then only in a relatively minor way. His interest in canals, with the emphasis in his area on the movement of coal by canal, went hand in hand with his foray into mining. At the time he was also, of course, busy with several other projects, especially Enclosures and drainage.

In the *General View..* Billingsley claims the following minerals were mined in Somerset in his day: lead, copper, iron, lapis calamaris, manganese, coal, lime-stone, paving-stone, tiling-stone, free-stone, fuller's earth, marl and ochre.*[1] This is an impressive list. In commercial terms, though, it came down to four: coal, lead, calamine and ochre, plus the quarrying of stone – the first three of these then being significant for employment and the economy in the area. Ochre, used as a colourant, was then mined on Mendip as a relatively small-scale specialist affair. Stone was then quarried mostly for local use; of this list it is now by far the most important, exported nationwide.

Landowners would invariably reserve all below-ground mineral resources from a lease and treat them separately to anything above ground. Much of the land Billingsley bought or leased was therefore subject to the usual clause reserving the rights to underground minerals, meaning he would not have benefitted from its extraction. In any case, no mineral deposits were thought to lie beneath his Ashwick land. In contrast, some of the land he bought on the Mendip plateau came with the rights to mining and did have mineral deposits beneath it, which he took advantage of.

Taking coal first, the North Somerset Coalfield stretched from south

---

* Very small amounts of silver are found in association with the lead, mostly around Charterhouse (where Billingsley did not hold land). The romans mined this and even minted coins of it, but it has not been important since.

of Pensford to just over a mile from Billingsley's home in Ashwick, the mines being in two groups – northern and southern (there was also a separate area of coal mines round Bristol). Coal mining has a very long history in the district. He will have been familiar with 'cole pitts' from childhood: coal occurs near the surface and had been mined close to his home since roman times, especially in the Nettlebridge valley on the Ashwick border.

By the seventeenth and early eighteenth centuries coal was being worked in deeper mines in several places near Ashwick. The surface of the land covering mine workings was subject to the usual annual rent, while underground minerals were subject to a share of the profits of extraction: in the case of coal, either one-eighth or one-tenth of the coal dug, or one-eighth or one-tenth of its sale value (usually the former in each case), payable to the land owner. The lord of the manor and some local families who owned rights could do very well from the mines, but as ever the actual miners were skilled labourers who merely scraped a living. But for the leaseholders and shareholders, apart from the risk, even when a pit was operative it was not necessarily lucrative. Billingsley judged that profits in the coal industry at that time 'were considerable for a few works, but small for the remainder – that on the whole they were not equal to the extent and risk of the adventure'.

He says there were 26 mines in the northern group, selling excellent coal at 5d a bushel (9 gallons), and that they then raised coal from 70 or 80 fathoms deep. Coal from the southern group (including Moorwood in Ashwick) was not quite as good, but made very good coke which could be used in forges or for drying malt (so will have been used by the Oakhill Brewery for the latter purpose). Because its weight and bulk led to transport difficulties, when roads were used the cost of coal was always closely tied to the distance from the mine: the price in some places is said to have doubled for every five miles from the pit. The coming of canals changed the price significantly.

There was a strong relationship between canal building and mining of all kinds; canals were always most desirable in areas where heavy goods needed moving. While the northern group of mines, including High Littleton and Camerton, fared well from the opening of the Somerset Coal Canal, the southern group, centred on Radstock (on the uncompleted branch), did not do so well, needing tramways to move their coal. Further south again, mines in the Nettlebridge valley were served even less well. Even if the uncompleted branch had opened this group would not have gained much from the coal canal. They would have benefitted substantially from the Dorset and Somerset Canal – had that project not also been abandoned.

The Nettlebridge valley coal works, therefore, still had no transport links other than the road network well into the nineteenth century. Added to that, the coal seams were very narrow and not as plentiful as in other places, so while this area has one of the longest traditions of coal mining many of the small mines were barely profitable by Billingsley's day. The Ashwick mines, closest to his home, including Moorwood and Pitcot, were small affairs, with many thin layers of coal interspersed between other rock, and difficult to mine. He will have been aware of their limited profitability. During his lifetime some of the mines a bit further away, near Midsomer Norton for example, became more profitable through going deeper. He reports that they had recently used steam engines – in this case not to pump out water but to raise the coal; but they now relied on water wheels moved by horses, so saving coal.[2]

It was the Radstock mines that seemed to attract him most. Major seams of coal were discovered there in 1763; mining started shortly after this (although the most lucrative mines in this part of the coalfield did not open until after his death). During the eighteenth century the Radstock mines operated under leases and issued shares: it was not until the 1840s that the Waldegraves took over the leases, as lords of the manor of Radstock and owners of most of the freehold there.* They will have already gained the one-eighth royalty from the outset.

As mentioned in chapter 10 it is very likely that – despite his dismissal of likely profits – Billingsley had a financial interest in coal before the start of negotiations to build the Somerset Coal Canal (1792). Three mines had opened in Radstock by this date: Radstock Old Pit (pre-1779), Middle Pit (1779) and Ludlow's (1782).[3] It is possible that Billingsley invested in one or more of these, though we cannot be certain as no records of early leases or shares can be found for these pits.† His only verified investment was in one small venture as late as 1797, on the outskirts of the Radstock group (near Clandown), at a time when the first three pits were doing well and prospects looked good.

A lease to Smallcombe Colliery (north-west of Radstock) was drawn up 22 May 1797, with Robert Tudway and Rev Richard Chaple Whalley leasing all the coal under 32 acres of land to twelve men.‡[4] The twelve included Billingsley, who invested in 'two undivided shares of one-sixteenth each'; there

---

* The Duchy of Cornwall and the Beauchamp family also had mines round the edges of Radstock.

† It is also possible, of course, that any earlier investment was in another part of the coalfield.

‡ Several of the men will have been well known to Billingsley, for example the Whalleys, who were prominent in the district: Francis Whalley served as an Enclosure commissioner with him.

is no statement as to how much these shares cost.* The next year Richard Whalley decided to become a partner, so articles of partnership were drawn up under a new name: the Smallcombe Coal Company. The pit was probably sunk the year after that, in 1799. By 1804 it was served by a tramroad branch from the line to neighbouring Clandown Pit. These two collieries were never directly served by the coal canal, but the movement of coal will have been helped by the tramway which linked them to it. 'Engines for drawing coal and water' are first mentioned for this pit in 1819. Down and Warrington reproduce a sketch taken from an old plan which appears to date from about 1835, showing the possible route of the tramroad, plus the position of the beam winding and pumping engine houses (so the pit must then have had two engines). It is not known how successful the pit was, or what sort of return Billingsley had from his investment. His executors sold his shares in 1818 for £300.† While the tramroad operated within his lifetime, installation of the engine was almost certainly after his death so he can have had no hand in it. From 1847 the Waldegrave family took over the working of the mine from the partnership – but again, well after his death.‡ By then the pit was almost exhausted and soon closed.

Billingsley was also involved with the Radstock mines through his work as steward and agent to the Waldegrave family (see chapter 16). In the late 1790s he helped negotiate the purchase of Richard Jenkins' estate on the western side of Radstock for the Waldegraves – at the 'express request and direction' of the Dowager Countess Waldegrave. This is believed to have included land covering coal deposits which they later utilised.[5] According to Gould (1999) three of the four working pits in Radstock had pumping engines installed in the first years of the 1800s.[6] It seems likely that one or other of these is the origin of Billingsley's statement that he had helped the Waldegraves avoid 'unforeseen calamities' in a mine which suffered 'ingress of water'; and that through his exertions a solution was found.[7] Gould mentions James Watt's patent as being infringed by the steam engine in Middle Pit which had been added in 1784 (rather too early to have been overseen by Billingsley in his

---

* Bulley [1953] says 'it would appear that the unit of capital required for a colliery undertaking in Somerset between 1760 and 1830 was within the £2,500 and £10,000 range', which suggests Billingsley's likely outlay was in the hundreds rather than thousands.

† This seems surprisingly low, but re-sale was limited to those already within the group (apart from specific exemptions), which must have affected the price. By comparing it with the re-sale of other shares at this time it seems to have been about average. It does not seem to signify that the mine was unproductive. Plus it was a difficult time to get investment.

‡ Interestingly, when the mine was dug the landowner was not Waldegrave.

work for Waldegrave) – this was replaced in 1801, and in 1804 the first beam winding engine was erected there. Perhaps this was the pit that had flooded, and Billingsley advised on the replacement – he is known to have taken a great interest in water and flood control, seen especially in his drainage of the Levels (see chapter 12).

As Waldegrave's agent Billingsley also dealt with James Walker. The latter had superintended the coal works for over twelve years, made account of the coal landed and every six weeks hosted the coal managers' meetings at his inn in Radstock. For this work Billingsley made him an allowance of £250, which he felt was 'moderate for his expenses'. This was the subject of a serious disagreement with Waldegrave (explained further in chapter 16).

Moving on, 'the Mendip hills are … famous for their mines' says Billingsley in the *General View.*, referring here not to coal but to lead and calamine mining, well established there in roman times and probably earlier.[8] Lead occurs in a relatively small area of about 17km by 8km, with underlying limestone rock. Almost all the ore found lay within 50m of the surface, on the Mendip plateau to the west of Ashwick in the vicinity of Charterhouse and surrounding parishes, including Ubley where he owned the manor of Hazel and extending to East Harptree where he owned other land. But, he says, by his day these mines near the surface were 'nearly exhausted' and they could not go deeper by reason of water ingress.

Billingsley must have thought that lead mining was still likely to be profitable in the 1780s and 90s. For example, there is evidence that he exchanged land allocated to him under Chewton Mendip Enclosure Act for land with lead mining potential.[9] Billingsley wrote that, 'in all probability millions in value may remain concealed in the bowels of this mountain… it is probable that lead, like coal, is more valuable in the deep'.[10] He believed that by cutting an 'audit' (adit) for a distance of about five miles, costing, say, an estimated £100,000, the mines could be drained and transformed, 'with 1000 subscribers paying in £100 each', the project employing perhaps 100 workmen.*[11] Billingsley focussed here mostly on the financial side, not exploring the practical at all. The idea was never taken up, though according to expert opinion it would probably not have worked in any case.

The history and organisation of Mendip lead mining is very interesting: from the late 1400s it had its own unique code of law (set out in full in

---

* In a note added to this section, Mr Paget (a local landowner), agrees with Billingsley that an adit would be advantageous, saying that the possibility had been discussed for some years. So although Billingsley is usually credited with the suggestion of the adit, it may be that he simply took up an existing idea.

Billingsley's *General View..*; it repays reading).¹² It was a lucrative business, so inevitably there were disputes over ownership of the ore and its extraction. There were four 'Mineries' (also called 'Liberties'), each with a 'Lord Royal' (or 'Lord of the Soil'). Their names and owners changed over time, in Billingsley's day they were: the lord of Priddy minery (Bishop of Bath and Wells); the lord of Chewton (Earl Waldegrave); the lord of East Harptree (after 1797 this too belonged to Earl Waldegrave); and the lord of Charterhouse (Lord Bonvill).¹³ Each lord of the soil had to hold two courts annually, each with twelve men or more to agree on mining disputes. The lordship of East Harptree encompassed the common or waste land of Mendip within the parish and manor of East Harptree, together with portions of adjacent parishes where Billingsley held land, so that he was involved personally in management of the mining as an owner, as well as through being Waldegrave's steward.

Those wishing to mine lead applied to a lord royal via his 'Lead Reeve': a fee was required for a licence, plus one-tenth of the value of the ore raised – the 'Lead Lot'. Through this 'in the past many thousands have been paid to the See of Wells [for example] for the Wells share only' says Billingsley. The lord provided the washing, cleaning and smelting area for the minery. Most men worked alone, a few in small partnerships, leaving long grooves as scars on the surface – from which the miners became known as 'groovers'. At that time the end of the process left about 25% lead in the slag. Later mining techniques which centred on reprocessing the slag to extract more lead have obliterated most of the remains of the eighteenth century mining process, but some rare remains still exist at Charterhouse. The long period of time through which lead has been mined on Mendip, together with the sheer number of small mines, had left the area riddled with holes, and the land above covered in shallow pits and hummocks – known locally as 'gruffy ground'. Much, but not all, of this has now been cleared so that the land is again usable for agriculture (though any area still with significant lead waste continues to be potentially deadly to animals). A good deal of the removal was undertaken during Billingsley's day, as part of 'improving' the land. Today, only those who know what to look for are likely to see any sign of the mining.

Lead ore on Mendip is often found in association with zinc, as calamine ('lapis calamaris'), though the latter was not commonly mined until after the time lead mining began to become unprofitable. Calamine is a zinc ore, then used by the brass industry in Bristol. Mendip calamine was said to be the best in England. The mining areas were mainly in the western part of the lead area, with smaller mines extending east into the Harptrees. Calamine was valuable: Billingsley reports its average price in the 1790s as being £5 a ton. The ore was sometimes found within a yard of the surface and seldom deeper than 30 fathoms. He claims that 400 or 500 miners were then employed in its extraction. Conditions are said to have been very harsh for the miners and the mining villages were wretched, the workers having a very poor reputation as violent and lawless. Billingsley first ventured into calamine mining in 1803, on a small scale, paying £6 0s 11d in cash for 'a Freeshare for calamine from Green Ore Flat'.[14] At the time he worked as steward for Earl Waldegrave, the lord royal of East Harptree. Waldegrave will have earned the one-tenth 'Lord's Share' in whatever was mined from Billingsley's freeshare, the amount not recorded.

Billingsley may well have had other mining interests on Mendip, perhaps in lead, but details of only one other venture have been found. This came two years later. On 1 July 1805 he and three partners were granted a 21-year lease by Countess Waldegrave, allowing them to raise 'Lead ore, ocre and calamine' on the old enclosures land at East Harptree, and requiring them to pay 'the Lord's Share' to the Waldegraves of 3s in very 20s worth of ore raised. At that time the latter's accounts for their Mendip estate were made up by JS Tudor, who was also one of Billingsley's partners in this.* The accounts show money coming in from a variety of sources, with deposits from two lead reeves (presumably for Chewton and East Harptree) – and including payments from Billingsley himself for 'his part of a Freeshare for Groving at Mindery Batch' (in East Harptree), for which he paid 4s 10½d. This probably relates to the above lease. However, this mine became the source of a serious dispute, as according to the Waldegraves the partners deviated from the rules of good mining. In 1808 the Waldegraves were granted an injunction to stop them from further mining. It was claimed that when digging, as soon as they met the water table they abandoned the shaft and dug a new one. The Waldegraves thought they should have installed a 'Fire Engine, wheel or other means' of ridding the flooded shaft of water. Had they done so, 'a much larger quantity of ore might have been gotten therefrom', instead of which their 'wanton and injurious methods sinking pits unnecessarily' resulted in the said lands being 'greatly deteriorated in value'.[15] The irony here is that Billingsley was so alert to the usefulness of drainage, including the use of 'Fire Engines', such as the one he installed in another mine for the Waldegraves themselves. No doubt, therefore, there was some other reason that the partners did not mine deeper, for example the stability of the ground, or the fact that calamine was seldom found at much depth.

---

*   Records are available only for the years 1801-5.

The quantities of lead, calamine and ochre mined on Mendip in the late eighteenth century were relatively small even in comparison to the amount of coal mined locally, but they were more valuable and had a big impact on employment and livelihoods at the time, in an area where there was little else available. Mining of all these minerals decreased after the early 1800s, although in the nineteenth century there was a brief resurgence of the local lead industry, mainly due to new methods for recycling lead waste from earlier spoil heaps. By the beginning of the 1800s zinc ore had begun to be imported and prices fell thereafter. With these changes Billingsley's chances of profit on any ore remaining under his land decreased significantly and he showed no further interest in mining. In 1825, after his death, the duty payable on imported metals was cut by Parliament: for lead by one quarter and for zinc by a half. This reduction effectively ended the mining industry on Mendip.

12
WATER MANAGEMENT AND DRAINING THE LEVELS

WHILE MINING WAS not one of Billingsley's passions, water and its management certainly was. 'He drained the Levels!' was the claim of the Bath and West Society shortly after his death.[1] There were, of course, many others involved over the centuries, but he did play a very important part in it, his work beginning in the early 1790s. However, well before he became involved in the Levels he was already interested in seeking ways to use and control water in general. At the brewery, for example, there was an elaborate system of pipes, ponds and storage tanks, which Billingsley will surely have paid attention to and may well have helped to engineer, if not the initial layout then the later improvement. A little after this he was involved with both canals and water management in mines. But apart from the brewery his first recorded practical involvement in water control came in the mid-1780s, in his experiments with water-meadows.

Water-meadows had been present in England for centuries; in the south-west the practice is known to have been used from at least the 1600s.[2] They became really popular in Somerset during the 1780s, probably due to the dissemination of ideas via the Bath and West, where serious discussion and publication on the issue began in 1783.[3] Coincidentally Billingsley bought the manor of Stoke Lane in 1783, and must have seen one of its fields as ideal for his own experimentation.

Water-meadows were intended to give two benefits. First, to produce an 'early bite' for stock, through flooding which warmed the soil and protected it from frost in early spring, resulting in the growth of grass earlier than on land left without that attention. Second, to encourage a good crop of hay later in the season, through irrigating the land at a time when it might otherwise be too dry for good grass growth, hay being essential winter feed for stock. For both purposes silt naturally carried by the water served as crucial fertiliser for the grass.* There were two main methods; in the eighteenth century the simple sloping-meadow system called catchworks became very common on Exmoor for example, but was not much used on Mendip due to the relative lack of surface water.

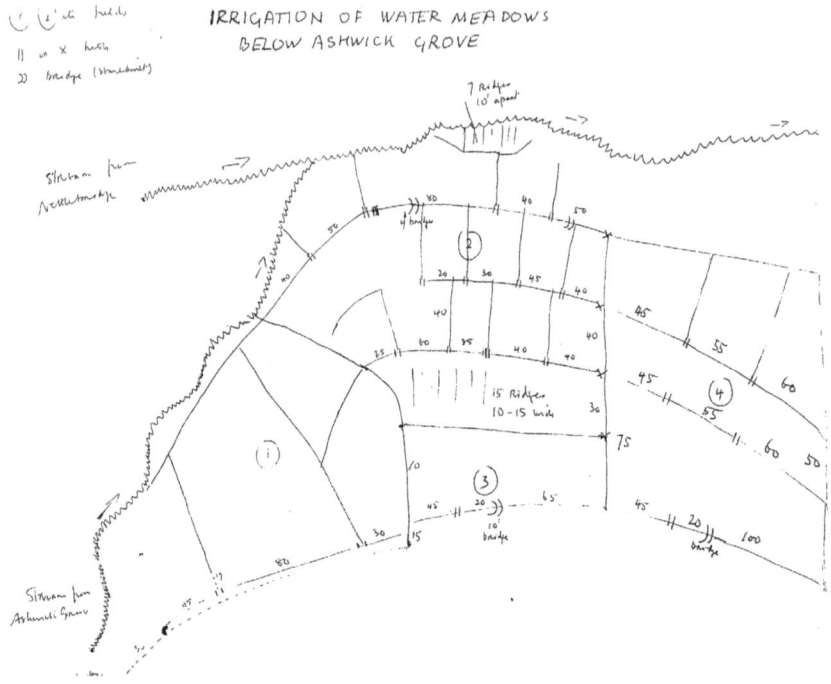

Water Meadow drainage system on land near Ashwick Grove
Downside Archaeological Society[4]

For his experiment Billingsley converted one of the fields on his manor of Stoke Lane land, close to Ashwick Grove. Unusually for Mendip a reliable stream ran through the small valley and past a sloping field.† The sketch map above showing its layout is hopefully self-explanatory. The noticeable downward slope on the land, towards the top of the picture, makes it easy to use the simple catchwork method. Water comes in from the Ashwick Grove stream at the field's highest point (bottom left of the sketch), runs through the various ditches, and is discharged into the conveniently flowing Nettlebridge stream. Although the system is long abandoned there are still some visible remains, including at least one of the stone hatches (to control the flow; they also allowed access across the ditches).

* Manure or other substances were also sometimes added to the water.
† Near Stoke Bottom Farm, between Nettlebridge in Ashwick and Stoke St Michael.

Water-meadows were then highly thought of especially as they allowed nutrients to be added, thus enriching the land. This would have been most important in upland regions with poor soil – although it did not particularly apply to the field in question, which was probably used as it was ideal for the experiment in its slope and access to water ingress and egress. There is no evidence as to how successful this particular arrangement was. The fact that, having engineered the system on his own land, Billingsley did not the publish any findings suggests that he found the idea was not as useful as he had hoped for his own area.

His personal involvement in water management thereafter was mainly confined to removing water, rather than adding it. However, he still considered the method itself to be useful: in the *General View..* he says that in the south-west district of Somerset 'watered meadows are invaluable'.*[5] He describes the water meadows of the Brendon and Quantock Hills as the best in the country, believing they should be used 'wherever water is of good quality and there is a possibility of conveying it'. He goes on to say with respect to the south-west district that 'most of the existing watered meadows lie on steep declivities; as the water passes quickly over them, and never lies stagnant'. He briefly describes the source and timing necessary for useful watering, but gives little detail as to the construction of the watering system. He also comments that 'the water issuing from the Mendip Hills is unfit for this purpose [of water meadows] carrying with it noxious particles destructive to vegetation'.† Maybe this is the reason he abandoned the water meadow on his own land.

Returning to Billingsley's involvement in the drainage of the Levels, he had lived all his life close to them and would have been very familiar with the fact that these potentially rich lands frequently flooded, making parts of them almost unusable.‡ As a fervent agricultural improver this alone would have interested him in working in the area. The Levels make up a substantial area of Somerset, extending to about 160,000 acres.§[6] The main part lies south of the Mendip Hills, forming a natural basin, bounded by higher land to the east

---

* All subsequent references to Billingsley in this chapter are from the *General View...* 2nd ed
† This comment is unexpected given that the Oakhill Brewery used spring water issuing from the Mendip Hills. Presumably he believed the offending particles were irrelevant in beer.
‡ Even today the Levels still floods at times, notably in 2013-4 when about 17,000 acres lay underwater for up to three months. In early 2023 some roads and large sections of the moor were underwater again, later that year there were more floods on the moor, through to spring 2024.
§ Much of the following section is indebted to Michael Williams (ref 7)

and Exmoor to the south, with the sea to the west.* Several rivers run through it, with a number of small slightly raised areas ('isles') between them. From north to south the major rivers are: the Axe, Brue, Cary, Parrett and Tone. The majority of the land is flat, much of it below sea level (by up to 10-15 feet at high tide.) †,‡ On the seaward side, around Bridgwater Bay, there is a narrow coastal clay belt which partially protects the land from flooding, but can also impede the flow of water away from it.

The reasons for flooding are threefold. First, the physical shape of the region – a basin surrounded by hills – encourages deposition of peat and silt through slow moving rivers; second, strong tides together with prevailing winds from the west, plus the shape of the bay, cause the sea to easily inundate the low-lying land; and third, there is exceptional rainfall in the area – especially on the surrounding hills which have a higher average annual rainfall than any other part of England. If high rainfall coincides with gales and/or high tides the water can overflow the usual river channels and flood the surrounding land – when it does it can be for long periods and with catastrophic effect. Despite previous efforts at drainage, in Billingsley's day large areas of the moor could still be underwater in winter, often for four or five months at a time, with occasional shorter but equally serious flash floods in summer.

Over the centuries an immense amount of work has gone into taming the waters. From efforts by the romans through to the end of the monastery era about 1,000 acres had been reclaimed, but storms and floods frequently ruined the work. Before the dissolution some river courses had been changed; sea-walls built; ditches dug (locally called 'rhynes', pronounced 'reens'); and sluices ('clyses') built at key outfall points across the rivers to control the water. For many years huge numbers of cattle, sheep and geese were fattened on these lands during the summer months. The area was also used for fishing and fowling; for collection of peat for fuel and reeds for thatching; and as pasture for hay and stock. But after the dissolution there was little further improvement for many years: the crown which had taken over the monasteries' land had acted to 'dredge, drain and reclaim' but these efforts were ultimately frustrated for a number of reasons, lack of maintenance an important one.

---

* There is also a small section north of Mendip, draining into the Congresbury Yeo, but there was not much serious attempt at its drainage until after 1810.
† The flatness of the land gives the area its present-day name, but it was not always called 'The Levels': in Billingsley's day and for some time thereafter it was more usually described as the 'Somerset Marshes', the 'Moors', or simply 'the Flat Lands'.
‡ About 18% of Somerset is below average tidal level, most of that on the Levels.

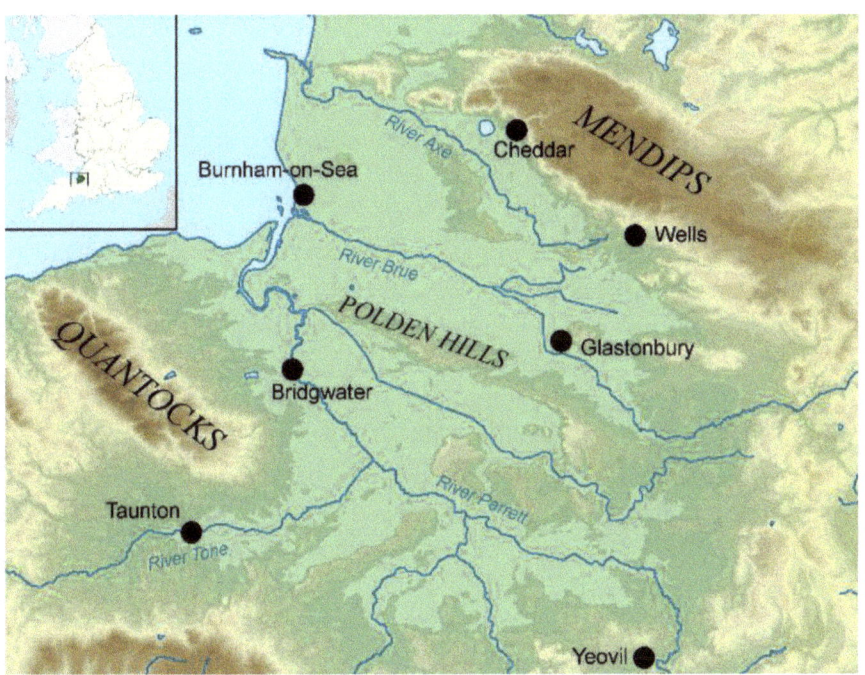

Map of the Somerset Levels[7]

The Court of Sewers, which had responsibility for maintenance, did not have sufficient powers to be effective and, crucially, was not empowered to undertake any new work. Sea walls and rhynes both needed frequent attention and repair, and there were arguments as to who was responsible for the repair, the money to pay for it, or the labour to do it. Many were reluctant to do their 'stint' if, for example, a repair made upstream was more beneficial to those downstream than to those who did the work. Some attempts at improvement actually made things worse, which did nothing to encourage action. The above account indicates very briefly just how much effort had already been made, but very little real progress occurred until at least the 1770s. Arthur Young called the state of the Levels at that time 'a disgrace to the whole nation'.[8]

By then the agricultural potential of this particular land had become glaringly evident – and increasingly so as the worsening economic situation meant that any suitable land was now at a premium, and most in Somerset had already been improved. While one of the major obstacles to improvement was the recurrent flooding, any scheme to prevent it was a real challenge within the technical capabilities of the day. In addition, agricultural improvers saw it as equally necessary to Enclose the land in order to maximise its profitability. And competing interests meant there were always objections – occasionally violent – to any scheme for either drainage or Enclosure. Some of the rivers, for example, were used for navigation (which was jealously protected) and for fisheries ('gurgites' or fish weirs often interfered with water flow making flooding worse). Other main local industries, such as peat cutting ('turbaries') and willow and reed cutting also resisted change. One of the most serious issues was that suitable grassland was in short supply, so any available grazing was grossly overstocked, resulting in stunted stock and reduced productivity. Landowners feared loss of rent, the numerous commoners feared loss of rights, so both were hostile – some very hostile – to change through either drainage or Enclosure. Several efforts at getting Enclosure acts through parliament failed.

From the late 1770s the agricultural improvers, including those of the Bath and West Society and later Billingsley himself, focussed on the area. Some local landowners had also become increasingly frustrated, so together they made a concerted effort to translate drainage and Enclosure schemes into acts. For the present discussion the relevant land can be divided into two main sections: first the Brue and Axe valleys, and second King's Sedgemoor.* The first area to be improved was the Brue in the 1770s, spreading to the Axe in the next decade. Billingsley saw the latter as having been much neglected because it was 'destitute of gentlemen's houses, probably on account of its stagnant waters and unwholesome air'.

At least 22 Enclosure acts in the Brue and Axe areas which had had land subject to waterlogging were passed between the late 1770s and 1810. Billingsley himself acted as a commissioner on the Levels only from 1791; he served in that capacity for five of the remaining seven acts in this period, these alone covering more than 22,000 acres.[9] In addition to already being an experienced Enclosure commissioner through his work on Mendip he seems to have also had a natural understanding of water control, producing innovative plans for its solution, with his advice being sought and acted on. All the acts mentioned here involved, in addition to the actual Enclosure work which proceeded in the usual fashion, the essential drainage – via rivers, rhynes, bridges and so on – so were a good deal more complex. It is noticeable that Billingsley gives much more attention to the process and reasoning behind the Enclosures on the Levels than he does for those on Mendip – an indication of not only the greater complexity but also his serious interest in water control.

The Levels Enclosure acts for which Billingsley was a commissioner are listed below.

---

* Billingsley was not involved in drainage or Enclosure in the area south of Sedgemoor.

Name	Date of Act	Date of Award	Extent in Acres
Rodney Stoke	1791	1793	600
Mark (Manor of E Mark)	1794	1797	150
Pilton and N Wootton	1794	1796	1,794
Butleigh	1795	1796	600
King's Sedgemoor	1791	1795	18,000
		Total acres enclosed	22,344 *

Table of Enclosure acts for the Somerset Levels for which Billingsley was a Commissioner [Billingsley was also a commissioner on two other Acts with land in the vicinity, already mentioned with respect to the Enclosure of Mendip, totalling a further 5,045 acres.]**

Billingsley calls the Axe and Brue section of the Levels 'Brent-Marsh', saying that by 1798 a total of 17,400 acres had been enclosed there, with 2,800 remaining as turf bog. These acts had brought some real and immediate benefits, with dramatically increased crops in some cases: he describes the results as 'astonishing'.

> The rhynes and ditches necessarily cut to divide the property, together with the deepening of the general outlets, discharge so much of the superfluous water, that many thousand acres which heretofore were overflown for months together, and of course of little or no value, are becoming fine grazing and dairy lands, 6 out of 7 parts are cleared of stagnant water, and rendered highly productive.

Whereas, he says, before the drainage and enclosure 10,000 sheep had rotted in one year in the parish of Mark. Unfortunately, this improvement did not last long. Following disastrous weather in 1799 these areas reverted to their previous condition – some crops, for example, were harvested by boat in that year. But the temporary gains had caused hope among landowners that with sufficient attention to drainage great benefits would occur. Further acts specifically for drainage were then seen as essential. While these were being debated a breach in the sea wall at Huntspill caused another serious flood. An unprecedented rate per acre was levied to pay for the repair. This

---

* This land on the Levels is almost double the amount of land Billingsley was involved with enclosing on Mendip - more than double if the two Wells acts mentioned immediately below that figure are also taken into account.

** St Cuthbert Out (East Horrington), Wells: 1792/4 and St Cuthbert Out (Mendip), Wells: 1793/5

was furiously objected to, but ultimately had the beneficial effect of finally bringing agreement in how to proceed.

Most of those involved realized only in retrospect that co-ordinated drainage needed to be in place *before* Enclosure, rather than after. Billingsley was an exception. By 1794 he had made clear his understanding of the flooding problem for the Axe, which had no barrier to the tide which flowed several miles up the river, and

> …choaks the lower parts of it with slime to such a degree that many thousand acres by the upper part of the river are… much injured. Were a barrier with proper sluices erected near the Bristol Channel, some of the most considerable windings of the river shortened, and the shallow parts deepened, not only the moors, but the old inclosures would be benefitted thereby, to the amount of at least £5,000 per annum.

The Brue drains a larger part of the marsh than the Axe. The former did have a barrier to the tide at the time, at Highbridge, but the tide could rise there by up to twenty feet, and as Billingsley says, the barrier's 'foundation, and the apron and cills of the sluices, are at such a height above low water that the drain is very imperfect'. The sluice (clyse) was insufficient in depth and also too far inland so that the mud and slime impeded the flow to such an extent that the current was not enough to dislodge it, resulting in excess water running many miles up-river. This took silt with it, in addition to that run off the hills around. He regarded the whole moor as having been formed by deposited mud, pointing out the number of trees, oaks, fir and willow, plus furze and nut trees, that have been found at depths of fifteen to twenty feet. The vale must have been formed this way, he says, 'as it is manifest that neither furze nor nut-trees will grow under water'. Without action silting would continue.

Billingsley suggested a 'Plan' for correcting the flooding problem, which can be most easily understood through the map drawn up for him by William White in 1794 (shown above).* Most importantly, he suggested that the Highbridge clyse should be rebuilt and enlarged, with the sill ten feet lower. The bed of the river should be reduced to an inclined plane of one foot in one mile from Meare down to Highbridge. Being aware that landowners would object as their land would be made too dry, he proposed hatches at various bridges, to control the flow. There would also need to be an entirely new cut dug twelve miles across the moor from near Dundon to join the Brue

---

* White was a highly skilled surveyor from Sand near Wells, who Billingsley held in high esteem – he was always his preferred cartographer.

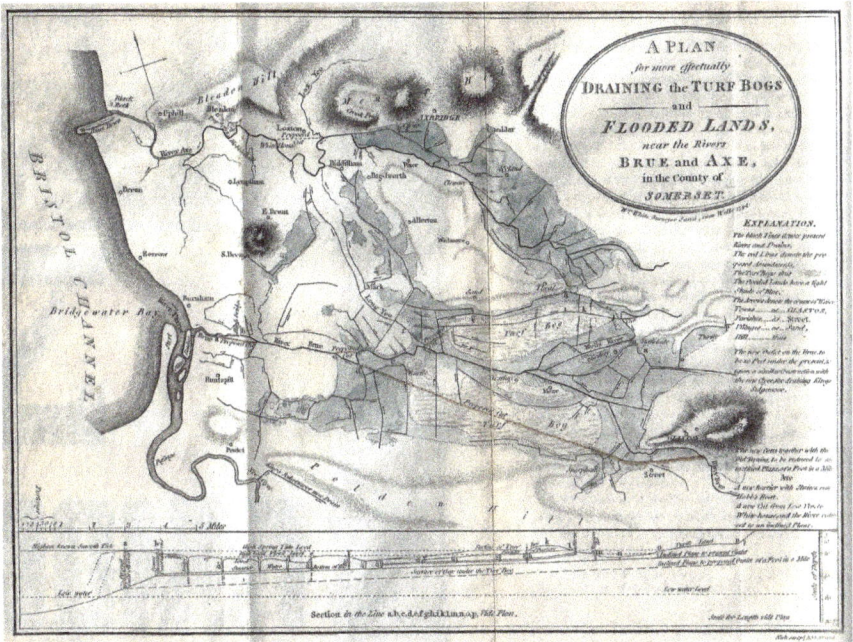

'A Plan more effective Draining the Turf Bogs and Flooded Lands, near the rivers Brue and Axe,' drawn up by William White in 1794 reproduced opposite p168 in the 2nd ed of the General View.[5]

lower down its fall – another innovative suggestion. These changes would mean a great improvement of the land, while 'the air would be rendered more healthful, and the exhalations of so large a body of stagnant water, which are wafted up to the high corn-lands of the Mendip Hills… would be unknown'. Finally, '*two* grand drains, accompanied with proper lateral ditches…would increase the rent of this district'.

Billingsley estimated the cost of his suggested scheme as being a total of £25,000 for the Brue area. In 1794 he thought the improvements affecting 9,000 acres of turf bog and another 5,000 flooded lands would together produce a profit of £231,250 over 25 years. By 1795 he had increased the area of flooded land he thought would benefit from the changes to 15,000 acres, and revised his estimate of profit to £356,250, with the same expenditure.* The lack of increase in cost he suggests is surprising as it was a time of serious inflation. He thought that expenditure in the Axe area would not exceed another £5,000, with similar benefits over a smaller area of land, but his estimate of benefits doubled between 1794 and 1795. He would have been

---

* These figures come from the *General View..*, 1st and 2nd editions, (pp166-173 in the latter).

well aware that these estimates would be very persuasive to any remaining objectors. The Brue Drainage Act was passed in 1802 and work completed in 1806; the Axe Drainage act passed in 1802 and completed in 1810.[10],[11] While Billingsley was not a commissioner on either of these acts nevertheless all his ideas were incorporated in them (apart from the line of one of the new drains).

Following the temporary improvements in the Brue and Axe valleys attempts had spread southwards into King's Sedgemoor (the second main section considered here, including the majority of the area between the Cary and Parrett rivers) to bring the land under improved cultivation. This land was in a particularly poor state: Arthur Young claimed that it lay at such a low level that the water 'had no way of going off it but by evaporation and in the winter it is a very sea'. It was unfit for cattle for more than two or three months in a year. Several attempts were made in the 1770s and 80s to put forward acts for Enclosure on the Levels, most ending in failure.*

Among those suggesting improvements was a local surveyor named Richard Locke of Burnham – an interesting character who bears some attention.† A near contemporary of Billingsley's, in many respects they had much in common. Locke was also an agricultural improver, who had been successfully experimenting for some time on land he had inherited at Burnham (on the clay belt). He knew the land intimately. He stressed the importance of drainage as well as enclosure and, encouraged by some landowners, began to produce his own plans for improvement in the 1780s, some years before Billingsley became involved. He claimed to have surveyed 20,000 acres of land for this purpose, but not without some problems. As an example of local antipathy to change, he says that during the surveying he was attacked: 'stoned, bruised and beat by the mob till the blood has issue from my nose, mouth and ears'.‡ Locke's plan for King's Sedgemoor, made in the 1780s, had involved cleaning and widening existing drains and several new, mainly short cuts, across the southern part of the moor. He estimated the total cost for this as being as little as £2,231.§ While not taken up at the time, many of his 1780s

---

* One act was thrown out of parliament when it became apparent the instigators were led by Lord Bolingbroke, who was (fraudulently) trying to use the act to recoup his gambling losses.

† Locke (1737-1806), published several important articles in the Bath and West's *Letters and Papers*. He also prepared a *History of Somerset* which he abandoned after Collinson published his.

‡ Locke was not the only one to suffer physical abuse for their efforts: in 1775 it was reported that William Fairchild was 'threatened with his life; and they went so far as to dig his grave'.[12]

§ He also proposed to make the waterway navigable as far as Greylake Fosse, at a cost of

proposals were the same as those which were incorporated into the 1790s plan, described below. Locke should have more general credit for his vision as a drainage engineer. Billingsley evidently approved his approach as he called on him for information and quoted him in the *General View..*, recommending him to the Board of Agriculture.

In 1787 a scheme for King's Sedgemoor, promoted by Lords Ilchester, Waldegrave and others, was prepared but was resisted and put off. Eventually, Sir Philip Hales, who had encountered 'much abuse and opposition' at meetings, nevertheless 'pursued the idea with great assiduity' and managed to put an act to parliament in the 1790 session. Billingsley describes the events of these years in some detail, to demonstrate, he says, that public meetings are not the best way forward when resistance is likely to be encountered. He believes it must be done by 'private application'. Given the objections by both landowners and commoners getting an act passed was not simply a problem of addressing water management and partitioning: any progress also needed to involve a diplomatic exercise. Fortunately, one of Billingsley's strengths lay in negotiation, a necessary skill in this case. An 'Act for Draining and Dividing a Certain Moor called King's Sedgemoor' was finally passed in 1791.[13] The award for this act was finalised in 1796 after the work had been completed and responsibility passed to other authorities, including, for drainage matters, the Court of Sewers.

The King's Sedgemoor act took in a huge area of some 18,000 acres, in 30 named parishes and hamlets. There were four commissioners, Billingsley being one of them. Rather than having a separate drainage act for King's Sedgemoor (called by Billingsley 'South-Marsh') as had been the case for the Axe and Brue, this act incorporated all the necessary over-arching drainage. The bill contained the usual clauses, that commissioners would be responsible for draining, fencing and enclosing the land, plus constructing roads and so on, together with powers to purchase any land needed and to sell land to defray expenses. Powers were also given them to clean all existing waterways and construct new ones, with new engines, sluices, flood gates and bridges specifically mentioned – work to accomplish the necessary drainage was included in the costs. Although all the usual Enclosure work was included in the main drainage act, about half the 31 parishes involved quickly obtained acts to complete the legal side of Enclosure, a few did so at a later date, but little is known of arrangements for the remainder except that Enclosure did somehow occur.

£3,255, an idea not taken up.

There had been much discussion – and much argument – over several years as to which plan would be most effective. In the *General View*., Billingsley describes the drainage process finally agreed on. The moor then drained into the river Parrett, some miles above Bridgwater – its fall 'trifling'. It was first suggested that all that was needed was opening and widening the old outlets and attention given to old clyses that no longer worked properly. The commissioners thought this would be obviously insufficient. Mr White the surveyor was instructed by the commissioners to find a better route for the main quantity of water, working on the problem together with Billingsley. The discharge needed to be lower down the river, at the 'old Dunbald-Clize'. It seemed a fall of ten feet could be gained. From the inland end of the moor to the outlet a drain would be needed, a distance of 10 miles, and a fall of 19 ft: 1" in ½ a mile. 2½ miles of this would have to be cut 15ft deep through the coastal belt. This was the main 'King's Sedgemoor Drain'. A secondary '18 Foot Drain' was also needed. This work would be very expensive. Plus, the river Cary would need to be diverted. There was some disagreement, but when Mr Jessop the engineer was asked to comment he agreed with their ideas.* So a sluice was erected and a channel 15ft deep, 10ft wide at the bottom and 55ft wide at the top was decided on.

According to Billingsley the whole plan produced a 'deluge of ridicule', critics claiming that the work of commissioners allowed three guineas a day would never be finished until the expenses equalled the value of the moor! But 'uninfluenced by letters or by menaces the Commissioners persevered'. Opponents objected that the drain only needed to be 26ft at the top – which would have led to collapses. Even at 55ft there were many slides, and these slides might continue, Billingsley thought, for years to come, the soil being 'so soft and morassy'. And expensive as this was, yet when finished he claimed that 'the probable improved value cannot be estimated at less than £450,000'. Having whetted the appetite of the investor he makes sure to give his oft-repeated reminder of the danger of overstocking: that all land becomes unproductive as a consequence of overstocking – 'then of what value is the land?'.

However, there were more problems to come. The land owner refused to sell land at the preferred site of the outfall, so Dunball clyse had to be rebuilt further inland. Poor sub-soil conditions meant that the foundations had to be built 4ft too high, which led to sedimentation. Many other clyses and bridges were erected. It was an imaginative and, for its day, extremely effective solution to the problem. The total cost of the act was £31,624, of which £4,314 was

---

* The engineer who actually worked on the project was Robert Anstice.

due to the commissioners and £15,418 paid for the engineering and drainage itself. 12,000 acres were affected, the expected increase in value for these being £525,000 over 25 years, and another 4,000 acres expected to increase by £50,000. Billingsley gives the total projected profit of the undertaking as £575,000 over 25 years. He remarks that there are also other lands in the catchment in the process of being Enclosed (including West Sedgemoor and North Curry) also benefitting from this work.

By a happy chance Billingsley's own copy of his work as a commissioner is preserved.[14] In the flyleaf he notes that he began the commission on 20 June, 1791 and ended it on 20 September, 1795. The Enclosure work was onerous and exacting: for each claim the landowner and occupier were noted, with acreage; quality and value of the land; the claim examined as to its legality; a decision made; and for each parish the expense of the process noted.* By March 1792 the clerk was able to advertise that lists would be displayed at key places in the district (mainly public houses, as was then common for such purposes) for persons to appeal against decisions as they thought fit. In the event, from a total of 4,063 claims only 1,798 were allowed. Given the amount of work the time scale is astonishingly fast, especially taking into account commissioners' other duties during this period. It is not clear how much of the work Billingsley did himself (no assistant is mentioned), but at the time he was engaged as a commissioner on at least seven other acts, preparing the *General View..*; renovating his home at Ashwick Grove; working for Earl Waldegrave and chairing various committees – and so on. Plus, he lived at a distance. The workbook carries no details about any drainage activity – only division and allotment. Sections of land in each parish were identified to help defray expenses, then auctioned in April 1795.†

Meanwhile, several other areas of the moor (not within Billingsley's purview) were still encountering objections – the following is just one example as an illustration of the difficulties. In a letter of 18 September, 1794, Canon John Turner of Wells Cathedral wrote concerning plans to Enclose: it was the 'earnest wish of the Chapter of Wells and of the Lords of Manors with the parish of North Curry to enclose the waste'.‡ However, upwards of 1,300 acres

---

* There were, for example, 107 separate rights to land in Othery, the land being of an average value of £2 4s 0 ¼d. 228 individual claims were made, 107 allowed (34 disallowed and 87 duplicates), the final cost to the Commission being £1,819. His investigation of claims alone took him more than sixteen months.

† 45 acres were auctioned from Othery, for example.

‡ There is no name for the addressee on the document ('Dear Sir'), but it was probably to Sir John Hippisley-Coxe, MP of Wells.

lie within the view of Lady Chatham of Burton Pynsent, so she opposes the business and:

> for the sake of her Prospect we are threatened with the weight of ministerial influence…the lands at present are not worth more than 6s an acre, but if enclosed would, as Mr Billingsley and others aver, be worth upwards of 30s an acre. The whole Commons wish to enclose 3,000 acres, so the difference would be near £8,000 a year… [in profit lost] so as not to offend her Ladyship's nice Eyes.*15

Turner suggests that if they sought to enclose only 1,700 acres – the part not visible from the house – and if he was given a letter to Sir Francis Basset, at present in Bath, he could then speak to Mr Elliott who in turn might speak to Mr Pitt… Mr Pitt, of course, being Prime Minister and Lady Chatham his mother. No records have been found as to the result of this machination, but the Enclosure act including North Curry was delayed until 1800.16

Several further acts, including one for Butleigh (600 acres), were passed in 1796. On 1 January of that year Billingsley wrote to accept the offer to serve as commissioner for this act, and:

> provided the Act is passed in March, I think it may be possible to subdivide and give possession of all the allotments so that Interested parties may have summer occupancy – which would constitute gain equal to expenses of the Act.†

More evidence that Billingsley's optimistic, swift and practical approach was unusually effective.

The Levels Enclosure acts in general seem to have had a very positive response, even though there had been so much more difficulty (over a long period of time) in getting them passed. The Level's land owners and small farmers (including auster tenants) must have quickly seen clear benefit from the Enclosure and drainage and probably felt relief that the acts were finally through. But commoners who had lost their claims were denied their previous rights, just as on Mendip. Judging from the figures there were large numbers of unsuccessful – presumably disgruntled – claimants. Perhaps the energy of objectors was dissipated before the acts were passed as there does not seem to

---

* The 'whole commons' suggests that either antipathy to Enclosure is waning or Canon Turner is exaggerating.
† When ready the clerk would send the draft 'to Mr Billingsley by the fish cart which passes through his parish'.

have been the anything like the level of discontent evident on Mendip. There would have been continuing work for labourers in maintaining the drainage routes, which may have helped somewhat. There does not seem to have been the same personal antipathy towards Billingsley from any quarter, in fact he is reported as having been praised for his work.

In the years following the Enclosure the King's Sedgemoor land was hugely productive, but despite the efforts made the continuing inadequacy of maintenance of the drainage system, plus repeated ploughing affecting the soil, meant that, like the Axe and Brue areas, Kings Sedgemoor became waterlogged again. By the 1830s it had become impossible to pasture cattle. Sheep, being lighter, were initially more attractive but suffered foot rot so root crops were favoured – only harvestable, though, in good years. Ever since, despite further efforts and much improvement, there have been sporadic on-going problems with flooding and urgent calls for better drainage. Recently environmental decisions have competed with and sometimes overridden the need for drainage or changed its focus in order to protect and repair the very important, and vulnerable, ecosystems. Meanwhile, the flooding of farmland and buildings continues.

13
Ashwick Grove

Ashwick Grove was Billingsley's home, but it was not until the 1790s that he was able to own it outright and transform it into the mansion he aspired to. Whether he simply renovated the existing building or built a totally new house on the same site has long been the subject of debate. The site itself had definitely been occupied for more than a century before he was born. It has a complicated history.

For a general description of Ashwick Grove in Billingsley's time one cannot do better than quote the Rev John Collinson. Writing in 1791 he says the house is:

> new built and in a very romantic situation. It fronts the south and lies in a beautiful vale where the inequalities of surface in front of the House are planted with great taste with firs and flowering shrubs. Behind the House is a fine wooded Hill scar'd with Rocks of Limestone rising almost perpendicularly from the foundation and effectually screen it from the North and East winds. From the East end of the House a deep narrow vale winds between two high Hills for several miles in which flows a pretty Brook in which are plenty of Trout. This vale is a most beautifully picturesque scene; the slopes of the Hills on each side being very steep finely wooded with a variety of Trees and shrubs and scar'd with prodigious Rocks which project thro' the foliage from the Lofty brow of the Cliffs...' [1]

It is interesting that this description hardly mentions the house itself – Collinson only saying that it is 'new built' – concentrating instead on its surroundings and the landscaped 'wilderness' which was fashionable at the time. Evidently the grounds conformed well to the prevailing taste. In 1791, the previous 'humble and retired' house, already inhabited by the Billingsley family for about 100 years, became a mansion befitting a gentleman of substance.

ASHWICK GROVE

Ashwick Grove in 1937²

The house lies towards the top right-hand corner of Fortescue's 1791 map (below), adjoining the garden labelled no 6. What is not apparent from the map is the depth of the valley. From the north-east edge of field no 4 the land falls away very steeply, suddenly dropping about 40 metres into the narrow valley in which the house lies. The other side of the valley is even steeper, more a cliff than a slope. The road running from the bottom centre of the map is the roman 'Fosse way'; towards the top it loops round (near the field named as belonging to Billingsley) where it drops down into the valley – even the romans had to use diversions to cope with very steep gradients. It is the shape of the site, protected from the prevailing south-westerlies as well as winds from other directions and opening into the valley down which any frost can roll, which gives Ashwick Grove its unique micro-climate, in what is otherwise a very exposed area.

The early history of Ashwick Grove and the surrounding land is fiendishly difficult to tease out, the result leaving many questions. The records go back well before 1700, but many are missing or undated. Briefly, the house itself lay within the manor of Ashwick, then owned by the Fortescues.* It is said that soon after the Rev Nicholas Billingsley (Billingsley's grandfather)

---

* As previously mentioned, the property lay on the boundaries of three parishes: Ashwick, Shepton Mallet and Stoke Lane (now called Stoke St Michael). Hence there was often confusion as to which parish it was actually in. While the house itself was in Ashwick, most of the land, even the garden, was in Shepton, with some in Stoke Lane.

Detail of Map of the Manor of Ashwick belonging to Earl Fortescue, 1791   the top right hand corner points roughly north; Ashwick Grove lies just south east of the 'T' for '[M]allet; the land to be sold in 1791 is shaded[3]

arrived in Ashwick, back in the late 1690s, he came to live at Ashwick Grove. However, the name originally associated with the whole property was 'The Fosse House', or 'The Fosse House Tenement' ('tenement' meaning land with a dwelling). This consisted of a 'messuage' (usually a small, undistinguished house) and about 22 acres, all the fields being pasture apart from Green's Wood. The property was in two parts, with two tenants. In an undated record, believed to be from the turn of the century, 'Mr [Rev] Nicholas Billingsley' is shown as renting two-thirds of the holding (plus other lands in Ashwick) and at least part of the house (see chapter 1).[4] The other tenant was a member of the extended James family, which the Rev Nicholas married into, probably soon after coming to Ashwick.* It cannot be confirmed, but seems likely, that Rev Nicholas built an additional house (or added an additional section onto the existing house) relatively soon after coming to live there, calling it Ashwick Grove.

By 1729 both the Rev Nicholas and the other tenant had died, so in 1731 two-thirds of the lease was awarded to his son Nicholas (Billingsley's

---

* There is no known marriage record for the Billingsleys, but it is likely to have been in 1698 or 9 (Rev Nicholas cannot have been in the district for long before that, and their first child, who died as an infant, was born in 1700). James' widow is named as a life on the Billingsleys' lease.

uncle), the other one-third to another (new) tenant.*[5] It is not definitely known which tenant cultivated which fields during this time. Following the Rev Nicholas' death and his elder son John senior's marriage in 1731 there may well have been changes in who lived where. John senior may have lived in Shepton Mallet, but is thought to have moved to Stoke Lane (Stoke St Michael) at some point.† Billingsley himself was not born until 1747. It has often been claimed that Billingsley was both born in and always lived at Ashwick Grove, though there is no conclusive evidence as to where he spent his childhood and youth.

Fortescue's 1763 survey says that at this date the property, no 1, was shared between two tenants, Messrs Billingsley and Burge.[6] A note adds that it 'is all very good Ground, and well worth the Sums set against it'. Nicholas also holds no 2 on the survey, described as a 'very good dwelling house'. This house appears to be Ashwick Grove itself. It is not clear where the 'new house' mentioned in no 1 is sited, or who lived in it (presumably Burge).

In a different version of the 1763 Survey, Nicholas Billingsley, given as then aged 59, and his brother John, given as 60 (actually 60 and 63 respectively), are named as the two lives for both no's 1 and 2. So despite the other tenant (Burge) being shown as holding part of no 1, the *tenancy* depended solely on the lives of the two brothers.‡ Unfortunately, they did not arrange a replacement third life, leaving it vulnerable in case of death, especially given their ages. Within a few years, in 1774, both Nicholas and his elder brother John senior died, in quick succession. Fortescue's Agent, William Miles, wrote to his Lord's secretary on 15 March, 1774: 'I am informed that Mr Nicholas Billingsley died in my absence [from home]. He was a Life on a small Estate now only held for the life of his Brother, a very old man'. We have only one side of the correspondence, but it is clear what happened. Billingsley himself, now aged 27, undertook the negotiations. Miles wrote to Fortescue again on 9 April:

> Mr Billingsley desires to add two Lives on a Messuage and Lands called Foss-House Tenement (No 1) now held by the Life of his father only, who I believe is near 70 years of age [he was 74], but of a very healthy strong Constitution, and also on a Dwelling House and Garden (No 2) held by the same Life. I cannot recommend to my Lord to renew the Estate but upon very advantageous terms

---

* Nicholas paid £32 for the tenancy.
† Fosse Farm in Stoke St Michael bears a date stone with the initials of him and his wife, built in 1759.
‡ At one time their mother Mary had been named as a third life, but she had died in 1756.

considering the age of the present Life, but the House if he can get a tolerable price I think it would be prudent to accept it at the present. The owner talks of improving it which is much wanted, and being a very old House, if it should happen soon to fall in hand I don't apprehend it would be of any great advantage…

Fortescue prevaricated. From his point of view delay would have seemed a good move as it looked as if Billingsley would have to pay more to buy a new lease than to add more lives. Unfortunately for Billingsley, before the issue had been settled John senior had also died (24 June), leading to the following from Miles to Fortescue (undated):

I am to give my Lord Joy of the possession of the Estate which was held by Mr Billingsley, who was last Friday seized by a Mortification in his Bowels and is since dead. I would recommend my Lord to sell again by Auction, which if he approves of, you will favor me with an Account of the lowest price his Lordship will accept both for the House and Estate, but as the House is a tolerable good one I would connect it with the Estate.

The condition of the house seems to have improved since Miles' letter of 9 April! Fortescue's position, though, was evidently not as strong as he would have liked, for on 13 August Miles writes again:

I have put this Estate up twice to Auction but cannot sell it for the price my Lord asks, the principle Reason is that Mr Billingsley being one of the Chiefs of the Presbyterian Brotherhood of which tribe almost the whole Parish consists, no person in the place chooses to bid against him, and as the Estate lies in the Parish of Shepton Mallett where the Poor are very numerous and burthensome, Strangers don't wish to be concerned with it, added to which too, on this last Account I cannot think it worth so much Money, having inspected the Poor Rates wherein I find it is charged at 15s 4d a Rate and this year there will be nine Rates collected amounting to £6 18s per ann besides Church Rates Landtax and Repairs, this makes a very material Difference which I daresay my Lord and Yourself will be sensible of.*

It was now Billingsley who prevaricated, holding out on the price for a new lease so that Fortescue did not get as much as he hoped. Fortescue must have reduced the price sufficiently. On 6 September, 1774 Miles wrote to Fortescue

* Note that 'selling' here refers to selling a lease to the holding, not the freehold.

once more: 'I saw Mr Billingsley yesterday and contracted with him of the Estate fallen in hand at 380L I am now preparing a Lease and shall send it by the Exeter Coach Friday next'. The front of the original 1731 lease was endorsed in 1774 with the names of three new lives: John Billingsley, 27, Thomas Parsons 17, and Isaac Parsons 5, sons of Isaac Parsons [Billingsley's nephews].

By this date, then, Billingsley had obtained the lease for the whole of the Fosse House Tenement and the 'very good Dwelling House' (Ashwick Grove itself), that is no's 1 and 2, including 22 acres, two houses and an allotment on Mendip of four acres.*

While the name 'Fosse House Tenement' was always used by Fortescue for both parts of the property in the period discussed above (there were several surveys during this time), the Billingsley family may have already been using the name 'Ashwick Grove' for the dwelling house (no 2). When writing about Billingsley's forebears retrospectively others always use that name for their home. The first known use of 'Ashwick Grove' by Billingsley himself is believed to have been 1781: he had used 'Shepton Mallet' as his address up to at least the end of 1779, but shortly after that date, in November 1781, he began to use 'Ashwick Grove'. Maybe he had moved into Ashwick Grove as a result of his marriage in 1779.

During the next few years Billingsley approached Fortescue several times in an effort to buy Ashwick Grove outright, but Fortescue always resisted. It was not until 1791, when Fortescue was already beginning to sell off his land in the district, that he succeeded in buying the freehold at a price he was willing to pay, seemingly by negotiation rather than auction.[7] As previously mentioned (in chapter 9), Billingsley paid a total of £46 10s for the whole of the Fosse House Tenement 'In Fee' – £41 for no 1 and a mere £5 10s for no 2, described as 'part of a dwelling and garden'. He already held the lease, with 82 years remaining. As a sitting tenant it was a relatively small sum to gain the freehold.

Having finally secured full ownership of the property Billingsley was now able to renovate it and lay out its surroundings to his own desires. The identity of the architect is not known, nor the builder or original plans. Although Collinson described the house at Ashwick Grove as 'new' in 1791, Robin Atthill, writing in the 1960s, claimed 'some parts are palpably older'.†[8]

---

* Miles' Accounts dated 23 Sept 1775 also mention payment of the heriot due at Nicholas' death, paid by Billingsley at £2 6s 6d, and that Billingsley had paid the £380 in full for the lease to two tenements and land at Ashwick.

† Atthill had been able to fully investigate the property prior to its partial demolition.

Internal evidence suggests that the 'new' house was built on the same site as the original, retaining some parts of it but with a larger footprint, which would mean the work was what we would now call a renovation – though with the house greatly increased in size and grandeur.

Much has been made of the fact that the building was set along the valley rather than across it, this being thought an odd choice. Presumably the direction of the sun is one reason it lies along – it allows light to flood the south-facing front elevation and main rooms even though it is in a narrow valley. More importantly, it allows the mansion to have a longer façade, making it look more impressive. Its position in the valley was the result of an earlier choice: it would have been very sheltered in an otherwise exposed location, though it also meant it had no 'prospect'. The house itself is cut into the cliff at the back, with stone rooms in the rock used for storage of perishables, as one would normally use a cellar (in more recent times these rooms have been separated from the rest of the house by a narrow driveway for access). It is not known whether the stone rooms were already present or were built at the time the house was being renovated. There are water tanks under the main house, rather than cellars, a constant water supply being problematic in those days.

It seems rather strange that after what must have been a large expenditure and a great deal of effort to complete the renovation, the Billingsleys moved to Bath relatively soon: 'I am now resident with my family at No 16 Argyle Buildings, Bath', he says in a letter of 1796.[9] Although this move appears to have been temporary and it was not long before he was back at Ashwick Grove, he moved to Bath again in the early 1800s for a longer period, settling at Camden Place (see chapter 17). He returned to Ashwick Grove finally in about 1808. After his death his wife and daughter stayed at the property only until 1814, when Marianne married.

It is not possible to describe the interior in Billingsley's day, in any case later occupants are known to have changed it significantly. The Strachey family (occupants from 1817-1940s) describe many different floor levels and something of a warren of small rooms inside, though how much of this dated from the original building or the renovation is hard to say.

There is a description of some of the changes the Stracheys made, written in the 1890s, but even from this it is difficult to imagine the original layout.[10] Externally, the house is typically Georgian, but, as Atthill remarks: 'from an architectural point of view Ashwick Grove was disappointing. The plain Georgian façade was diversified only by a rather curious apsidal porch – so that one appeared to enter the house at a corner.'[11] A coloured pen sketch of the house made by the diarist Rev John Skinner in the 1820s does not show the pillared portico on the front that appears at a later date. This must have been added in the Stracheys' time.[12] Skinner's sketch is the nearest record we have of the house in Billingsley's day. 'Without [the portico] the façade in Billingsley's day would have been plain indeed', says Atthill. Perhaps the 'disappointing' architecture he describes was the reason successive commentators seldom mention the house itself in any detail, always preferring to describe the grounds.* But other Strachey changes, including a new roofline and attic floor, plus a heavy cornice, make the simple frontage of Billingsley's day seem much more appealing than the grandiose structure created by the Stracheys.

All that remains of the house today is its shell. Sadly, the whole site of the Grove is now extremely neglected and in places dangerous, the mansion having been uninhabited since the late 1940s. Billingsley laid out the grounds at the time he renovated the house; despite the recent neglect they seem to a large extent to still retain the original plan. The estate was fashionably gentrified to include parkland with the required 'picturesque landscape' extending some distance from the house.† He began the scene on the western side by planting a stand of beech trees on one side of the approach to the house – probably mainly to provide extra screening and protection for the site from the weather, but also introducing the sense of a wild hilltop view, visible from the more practical drive as visitors enter. When the mansion was occupied the garden had well-tended lawns and shrubs near the house, but moving away from it through a gap in the garden wall and down the valley the style becomes more typical of the picturesque landscape which Billingsley evidently espoused.

Known as 'the pleasure grounds', this area still contains traces of a number of stone features, making use of the natural rock. There are flights of stone steps, a gothic arch in a wall, and a semicircle with a low wall surrounding a seating area. In the garden close to the house there are indications of the layout which must have existed even in Rev Nicholas' day. Reference has already been made (chapter 2) to the famous summer house where, as recorded by Collinson, Dr James Foster 'composed many of those excellent discourses on natural religion and social virtue… which have been read with universal admiration'. These words were inscribed on a stone by the summer house as a memorial to Foster. The stone is long gone, as is the summer house, but remnants of the steps that led up to it are still discernible among the undergrowth.

---

* A few elderly persons locally, who remembered the house as it was in the 1930s and 40s, said it was surprisingly light and sunny, with attractive main rooms downstairs. But the attics – where servants slept – were damp and dreary, leading the local doctor to tell at least one maid she would need to live elsewhere to regain her health.

† The parkland setting still remains at what is now Park Farm, once part of the estate.

Still within what were once the pleasure grounds and at some distance from the house, is a grotto. This stands partway up the side of the deep and rocky valley mentioned above, running through into ancient woodland (Harridge Wood). The grotto probably dates from the time of Rev Nicholas rather than his grandson.* It is primitive in construction – perhaps better described as rustic – and does not show the vision and fine craftmanship of, for example, the grotto complex at Pondsmead (only a mile away, see chapter 4). It is simply an igloo built of rough limestone blocks, the back of the grotto being formed by the rock-face. Where the private gardens ended there was a pair of wattle gates in Billingsley's day. Further again down the valley were ponds, originally made by damming the small stream, with watercress beds, and then what is now known as Keepers Cottage. Over the years this has been used for a variety of purposes: in Billingsley's time it is thought to have been an edge-tool mill.† Beyond that again are Billingsley's water meadows.

We do not know which other surviving garden features date from the Rev Nicholas' time, but we do know Billingsley was very interested in the landscape and garden around his home. He had applied to Kew for advice on appointing a gardener, prior to renting out his home in 1803, and it appears a young Irishman (unnamed) was recommended.‡ We have a very good idea of the state of the gardens at that time, the information coming from the lease Billingsley had drawn up when he went to live in Bath. The lease – a long and detailed document – was made on 22 October, 1803, for five years, between John Billingsley and Lydia White of Itching Ferry, near Southampton.¹³ It includes the house, together with stables and coach house, plantation, woods, lawn and pleasure grounds, as far as the 'wicket' (gate) eastward and a small meadow near the house – together being three acres, plus Green's Wood and a further ten acres. Also included was the cottage (lodge) near to the Bath

---

* Construction of a grotto may have been prompted by the fact that Foster knew the poet Pope, who had his own grotto built at Twickenham in 1720.

† It is now a bat refuge.

‡ In a letter of 30 Mar 1802 he thanked the official at Kew for taking trouble over this. It is one of very few letters from Billingsley to survive. Addressing the letter to Mr William Forsyth, (a well-known botanical figure in those days) he says the Irishman could come to Ashwick Grove at any time, adding instructions for the journey. He should take a coach to the Full Moon on the Bridge at Bath, then take 'the Shepton Mallet Caravan, which passes my house', giving precise details of the various elements of the cost of the journey. The terms of employment were: '2 guineas per month and 1 guinea over should he stay a whole year, and lodging but not washing; all my servants wash abroad' – presumably referring to clothes washing! [Reported in a letter from E.H.Ford to an unknown newspaper; from the archive of Robin Atthill.]

Road, the entrance way planted with fruit and other trees, together with 'all ways paths passages waters watercourses gates styles hedges ditches floures... advantages and appurtenances whatsoever to the said premises'.

It is noticeable that Billingsley was very concerned about the gardens and land, stipulating the way in which they should be maintained – 23 lines are given to the necessary maintenance of the grounds, and only 3 lines for maintaining the buildings as a whole, including the house. The following extract gives some idea of the flavour:

> Lydia White shall not cut down dig up lop top or shroud any tree or trees of any nature or kindsoever (except the usual and proper pruning of the wall and fruit trees) or any wood or underwood now growing ... shall keep the lawn well mowed and the borders walks and pleasure grounds and other parts of the premises in good order and repair during the said term and the stiles fences belonging ... near the walk from leaves or other things ... ornamental stone buildings or rock work...during the winter season weaving the same with straw ... keep in neat order and good repair.

Lydia White is to pay rent of £135 per annum – but no doubt, if she had honoured the lease, her outlay on gardeners would have been a substantial additional cost. The family was back in residence by the time the lease was due to end in 1808.

After Billingsley's death in 1811 his trustees tried to sell the estate, unsuccessfully for some time. In 1814 it was advertised in the *Bath Chronicle* as:

> Ashwick Grove, with the Pleasure Grounds, Plantations and Woods there adjoining, the property and residence of John Billingsley Esq, deceased. The House comprises a dining room, drawing room, breakfast room, housekeeper's room, servants' hall, cellars and other convenient offices, on the ground floor. 8 bedrooms on the first floor, and three servant's rooms. Coach-house for two carriages, harness-room, stabling for eight horses, and entrance lodge; excellent kitchen garden well stocked with fruit trees.[14]

The adjoining land comprised 20 acres plus a further 50 acres including the Fosse Estate, Park and two cottages. It did not sell. In 1815 it went to auction, but again did not sell. It was not until 1817 that Billingsley's executors managed to finalise a purchase. The Strachey family paid £5,500 for the house and about 20 acres, plus a further 20 acres of parkland and pasture including two cottages. Although during their time the Stracheys made a number of internal changes to the house they did not add to its footprint. The catalogue produced for the 1937 auction gives a good idea of its extent at that time, if not of its layout.[15] The 'Summary of Accommodation' shows a slight increase in accommodation, with 9 bedrooms, 4 bathrooms and 5 servant's bedrooms. The main rooms were large, in proportion to a stated ceiling height of 12ft downstairs and 9ft 6in upstairs.

The 'last Squire' of the three Strachey generations died in December 1936 without an heir; his widow left the house soon afterwards. Following this it was rented out for a few years, during which time it began to be stripped of any desirable items. During the Strachey's time the estate had been increased to 1,334 acres, spread over three parishes. Much of the land had in fact once belonged to Billingsley, but was not bought by the Stracheys at the time of their purchase. The last Strachey bequeathed the estate to the National Trust but they refused it. In 1937 the land and eight associated dairy farms were sold at auction, but once again the mansion itself proved difficult to sell.

Ashwick Grove 1963
the trees above the house had recently been felled and other vegetation cleared to assist the part-demolition[15]

In 1955 the house was sold for demolition, anything of any value plundered, and the site abandoned with many of the walls barely supported. Most of the grounds were disposed of and the ancient woodland stripped of its trees – for the price of the timber. It is now hardly recognisable as a house, with the garden features more visible than the walls of the mansion, which are covered in vegetation dripping with ferns and lichen, disappearing into the green surroundings.* The branches of the old yew trees meet above the path through the garden, so that looking up one feels as if in a cathedral. Badgers are thwarted only by the largest of the stone features as they dig their setts along the glade. The site remains in private ownership. Abandoned, it lies bereft.

---

* It is likely that it is within the local area now identified as temperate rainforest.

14
GENERAL VIEW OF AGRICULTURE

IN 1793, WHILE working on a myriad of other time-consuming projects and probably still finishing the renovation of Ashwick Grove, Billingsley was commissioned to write a book on the agriculture of Somerset. This proved to be the work he is best known for today – to give it its full title *'General View of Agriculture in the County of Somerset, with Observations on the Means of its Improvement'* (shortened here to the *General View..*).[1]

The *General View..* was initiated by the newly established Board of Agriculture as part of a series involving agricultural surveys of every county in the land. The board was the brainchild of Sir John Sinclair, 1st Baronet of Ulbster, (1754-1835). Sinclair was a Scottish landowner with a considerable estate, an active MP and writer and thinker on economics and agriculture. He was regarded by many as the outstanding figure in eighteenth century Scottish agriculture.* He held strong views on the need for agricultural improvement and an increase in food production, thinking this needed national action. During the early 1790s there were increasing economic difficulties: the supply of grain, and therefore the price of bread, was a particular issue. Sinclair argued that the country could produce sufficient grain provided farming and utilisation of land improved; this would enable imports to be blocked – thereby also protecting the interests of the landed class. Sinclair campaigned for two years for some means to encourage agricultural change before he got agreement from prime minister Pitt to set up his proposed board. The board was finally formed in 1793 as a chartered institution in receipt of a small government grant. It would receive £3,000 annually – but even its establishment and modest grant encountered on-going opposition from, among others, some in parliament and in the church.

---

* Sinclair's initially excellent reputation is now tarnished: historians now generally see him as being the main instigator in 'unleashing the tragedy of the Highland Clearances', with their wholesale expulsion of tenants by clan chiefs in the Scottish Highlands to make way for sheep.

Sinclair felt it necessary to first discover and document the current state of agriculture nationally, in terms of knowledge and results.[2] Despite the fact that the agriculturalist William Marshall had helped him in setting up the board and initiating the series of reports, it was Arthur Young who, due, it is said, to his contacts, was appointed as its first secretary (this, incidentally, was a major factor in the bitter rivalry between the two).* Sinclair aimed 'to reduce agricultural knowledge into a regular system, at least to ascertain what is already known and what is still wanting'; and to find ways in which the present state of agriculture in each district could be improved. It was expected that this information would point to measures which legislation might make to promote improvement; as part of this, the board saw encouraging Enclosure as one of its main concerns. This in turn would enable it to develop theories and systems of agriculture, including Sinclair's planned 'Code of Agriculture'. It was a very ambitious agenda: retrospectively Sinclair described it as 'the greatest undertaking ever attempted by an institution.†[3]

Each English county was to have its own survey. Sinclair had already begun commissioning individual *General View..* reports (very likely including Billingsley's) even before the board was actually established. Assembling the information for each survey was expected to take five or six weeks, which was a gross underestimate of the reality. Many of the authors gathered the information via a 'Tour' (as was common in those days), others used questionnaires, and/or lists of landowners and farmers to interview, though those using such methods tended to be authors who (unlike Billingsley) were not already sufficiently familiar with the locality. Billingsley was one of 90 writers chosen for the original surveys. His survey was completed very swiftly, especially considering his other commitments, being published in November 1794. It was one of the few 'Entered at Stationers Hall' under his own name: this meant that he held the copyright, not the board.

To understand the usefulness of the surveys it is helpful to consider briefly the progress of the whole set of reports before looking in more detail at the content of Billingsley's own volume. Each report was to follow

---

* William Marshall (1745-1818); practical farmer and agricultural writer. He thought at least twelve months observation of farming in an area was necessary before a realistic assessment could be made, this method of research differing from Young and others at the time, who mainly relied on tours and questionnaires (Marshall's approach on this may also have discouraged his appointment to the secretaryship as authors for the series would have been too difficult to find). Marshall's two main works were a twelve-volume study of England's *Rural Economy* and a *Review and Abstract* of the Board of Agriculture's County Reports.

† Due to difficulty in accessing the original data the following section relies heavily on Holmes [see References].

the same plan, set out by the board, the subject areas indicated in the table below. Most of these headings, such as land use and rotation of crops, were obviously agricultural; but others, such as the nature of leases and commerce and manufacturing for the district, were not, though they were relevant to agriculture's success. Authors would almost certainly be more expert in some areas than others. As shown in the table, discussion of possible improvements and obstacles to those improvements was also required, plus a list of those in the county who the board might usefully correspond with.[4] The books' margins were to be wide enough that readers could write notes in them, to be returned to the board for its use.

Subject	Subject	Subject
Soil and Climate	Seed Time and Harvest	Price of Provisions
Land Ownership	Inclosures	State of Roads
Occupation of Land	Advantages from Inclosing Land	State of Farmhouses and Offices
Land Use	Site and Nature of Inclosures	Nature of Leases
Grass Cultivation – Species of Stock – Status of Breeds	Impact of Inclosures in Population	Extent of Commerce or Manufacturing in the District
Watering of Land	Common Fields	Societies for the Improvement of Agriculture
Types of Grain Cultivated	Differences in Rent – Common Fields / Inclosures	Spirit of Improvement and its Excitement
Rotation of Crops	Extent of Waste Lands	Improvements to be Undertaken in Livestock and Husbandry
Fallowing	Draining of Land	Obstacles to Improvement
Ploughs, Carts and other Implements	Paring and Burning	The Most Active Farmers who Could Correspond with the Board
Use of Oxen and Horses	Woodlands	

Subject areas of the original Surveys commissioned by the Board of Agriculture, first phase of reports, 1793, derived from Holmes[4]

While Billingsley was easily chosen as the most appropriate author for Somerset, there was more difficulty in determining who should write some of the other reports, the board eventually deciding on several authors who were not truly agriculturalists. The quality of reports was therefore very variable, and some authors were slow, so that the first phase of county reports was not finished until 1798.[5]

Some reports were better received than others. Reaction to Billingsley's was positive. As a set they had a poor reception, but were, as Arthur Young felt, 'too severely criticised'. The speed with which the board expected to have reports completed was definitely a problem, several authors running out of time against the weather (in those days winter could mean travel was next to impossible in some places). The majority of reports, begun at the earliest in 1793, were received by late 1794 – an astonishing speed considering the amount of work involved. The authors were paid £100 per survey, but this 'renumeration [was] so inadequate [that] the whole allowance made to each surveyor did not exceed three months travelling expenses of a shopkeeper's rider' complained Young – who had written two of the initial reports himself.[6] Considering this, and the time and effort put into each volume, no doubt the authors would have been dismayed to find that the board, initially keen to get on with the printing and dissemination, soon began to distance themselves from the outcomes. Nevertheless, the reports were widely distributed and many replies and comments were returned to the board as intended.

Because of complaints, and the board's general dissatisfaction, already by 1795 a second phase was underway. The surveys were all to be rewritten according to a revised plan. The new model now laid down a clearer scheme, with subject matter to be kept in the correct sequence. There were to be seventeen chapter headings and numerous sub-headings, conclusions and appendices, the table below shows these in outline (without the many sub-headings). The stylised scheme meant that interested parties would be able to check methods and outcomes not just for their own county, but also easily compare them with other counties. Between 1795 and 1817 the 'corrected' or 'reprinted' surveys were published, with Sinclair now calling the first phase 'the rough draft of the survey', claiming that it had never been intended that they should be preserved.

Some of the first reports had been found so wanting, through being inadequate, incomplete or both, that many of the second phase had to be written by different people – in England only 13 of the 42 second reports had the same author and in Scotland as few as 4 of 23. Some of the authors were asked to rewrite their surveys for the second phase so that they adhered to the new scheme; Billingsley was one of these. Of others in the south-west region only Thomas Davis of Longleat was also asked to revise his.* A letter to Billingsley on 24 May, 1796 urged progress – the board was

> very desirous to hear that you are employed in new arranging and completing your report... The original met with such approbation as ought to encourage you in the undertaking with fresh spirits and renewed vigour... We have written to Mr Crocker of Frome to collect statistical information for you. He will follow your instructions in that part of the undertaking... The Board has elected you and Mr Davis corresponding members [of the Board] as a proof of our esteem for your talents.[7]

Chapter Number	Chapter Heading	Chapter Number	Chapter Heading
	Preliminary Observations	10	Woods and Plantations
1	Geographical State and Circumstances	11	Wastes
2	State of Property	12	Improvements
3	Buildings	13	Livestock
4	Mode of Occupation	14	Rural Economy
5	Implements	15	Political Economy – as connected with or affecting agriculture
6	Inclosing – Fences – Gates	16	Obstacles to Improvement
7	Arable land	17	Miscellaneous Observations
8	Grass		Conclusions
9	Gardens and Orchards		Appendices

Subject areas of the 2nd phase of Surveys by the Board of Agriculture, 1795 derived from Holmes [4]

However, once the second phase of reports were printed there was again some shifting of position on the part of the board, so that after the turn of the century they 'expressly disclaimed all responsibility as to the particular opinions advanced'. The project had turned out to be much more difficult than initially envisaged. In many cases authors who had replaced those

---

* Unfortunately, Davis died before finishing the manuscript, his son (also Thomas) completing it for him.

judged inadequate used their predecessors' work as a springboard to their own completion of the second edition, although others appear to have started from scratch. Billingsley's revised report, bearing the words 'Second Edition', was first published in 1795 (2,000 copies were printed in November of that year); it was reprinted twice by 1798.[8] It was 1810 before almost all revised reports were received. Despite the many failings and criticisms of the reports – and of the project as a whole – it was quite an achievement: extremely helpful to agriculturalists and others both at the time and later as the only full-ranging national survey of agriculture made during that period.

As a good proportion of the most important content of Billingsley's *General View..* is taken from or repeated in his other papers (several of which are discussed elsewhere in this book), its content is not considered here in detail; instead most of this section covers more general points. Like many other authors – and to the frustration of the board – in the first edition Billingsley did not follow the required plan as closely as he might, so that the reader needs to pay very close attention in reading the whole text to determine whether an item is included or not. There is no list of contents and no index. This is typical of Billingsley: his writing is knowledgeable, interesting and lively, but organisation of material is not one of his strengths, with him sometimes flitting between subjects without obvious reason. He can also be guilty of poor sequencing and lack of signposting. This is to a great extent remedied in the second edition, as the board proscribed not just the contents but also their sequence. Accordingly, as part of their instructions, the second edition begins with a list of contents, followed by an index, then Sinclair's introduction, incorporating the board's plan – making the whole thing much more user friendly. Sinclair's section shows his own superior ability in this respect. In addition to the new ordering and repackaging of content, Billingsley's second edition also differs from his first in that it 'contains considerable addition and amendments', accompanied (as stated on the title page) by 'the remarks of some respectable Gentlemen and Farmers in the County'. So we can assume he had received and acted on some useful feedback from the first edition. Billingsley also credits the fourteen Somerset 'Gentlemen' who gave him information to help in preparation of the survey – but adds that there were others who might have been useful but declined 'through suspicion [having] an ill-founded apprehension that the establishment of a Board of Agriculture was preparatory to additional taxation in some form or another'.

To return to the first edition, Billingsley begins with a short introduction, consistent with the 'Preliminary Observations' required by the plan; following this he announces that he will divide the county into three sections. First, the

GENERAL VIEW
OF THE
AGRICULTURE
OF THE
COUNTY of SOMERSET,
WITH
OBSERVATIONS
ON
THE MEANS OF ITS IMPROVEMENT.

DRAWN UP IN THE YEAR 1795,

FOR THE CONSIDERATION OF THE BOARD OF AGRICULTURE AND INTERNAL IMPROVEMENT.

BY

JOHN BILLINGSLEY, Efq;
Of ASHWICK-GROVE, near SHEPTON-MALLET.

AND NOW RE-PRINTED

With considerable ADDITIONS and AMENDMENTS,
Accompanied with the REMARKS of some respectable GENTLEMEN and FARMERS in the COUNTY.

———•••◦|◦◦◦◦◦•••———

SECOND EDITION.
———•••◦|◦◦◦◦◦•••———

In urbe luxuria creatur: Ex luxuria exiftat avaritia neceffe eft: Ex avaritia erumpit audacia; Inde omnia scelera ac maleficia gignuntur. Vita autem hæc rustica quam tu *agrestim* vocas, parsimoniæ, diligentiæ, justitiæ, magistra est.
                Tullii Orat. pro Sext. Roscio.

The City creates luxury; from luxury necessarily proceeds rapaciousness; and from rapaciousness breaks forth insolence: Thence are engendered all villainy and wicked deeds: But this country life, which you call clownish, is the regulator of œconomy, industry, and justice.

BATH,
PRINTED BY R. CRUTTWELL, FOR THE AUTHOR;
AND SOLD BY
C. DILLY, POULTRY, LONDON.
MDCCXCVIII.

Title page of Billingsley's General View... second edition, reprinted 1798[9]

'North-East District, from near Bristol on the west to Bath and Froome [sic] on the east', this section including Mendip; second, the 'Middle Division, from the boundary of Mendip in the north, Bridgwater Bay in the west to Chard on the south'; and third, the 'South West Division', incorporating the remainder. This is a useful and practical partition, especially as it divides the county into sub-regions with different environments and agricultural practice; it has been used by many authors since. He immediately moves into discussion of the first, north-east district. Initially he considers the coastal area, its need for drainage, and how this might be accomplished, moving on to coal-mining. But from here on the plan becomes less easy to follow, with no headings dividing the text until we come to 'Rotation of Crops' and the 'Size and Nature of Inclosures'. The next clear heading is 'Survey of the Mendip Hills'. Here he is on very firm ground, the great majority being a repeat (only slightly reworded) of what he has written elsewhere. Moving on to the middle division, much of his discussion, as might be expected, is of drainage, the need and methods – again, a subject well known to him. He begins with grazing management and dairy management, both with costs and outcomes of 'Brent Marsh' (lands round the Axe and Brue), then moves on to the 'South Marsh' (Kings Sedgemoor); the need for drainage and costs involved, with some attention to crops and produce towards the end. Finally, the south-western division is covered very briefly. Here he mentions only topics which he considers peculiar to the district, rather than being evident in other areas of Somerset: for example, the cultivation of rhubarb and the need for improvement of the considerable areas of 'swamp'.

While in the second edition Billingsley kept more closely to the main headings required, he still retains some irregularities which had been evident in the first: in both editions the three main divisions are very variable in quantity and in quality. Of the 199 pages in the first edition, almost half of the main part of the text covers the first district, two-thirds of that being on Mendip alone. Just a little less space is then given to the second district as to Mendip, half of this being on South Marsh (which he had just finished working on at the time of writing). These were of course the areas he knew best, so arguably his comments were most useful. The third section merited very few pages, in which there are several short subjects. While the second edition is considerably longer in total the proportion given to each section remains about the same, so is again unbalanced. In the first edition Billingsley had covered the board's various subject headings in one or more of these divisions in some detail, then only given them further attention in subsequent districts if he felt some important additional comment is necessary. If this were a stand-alone report

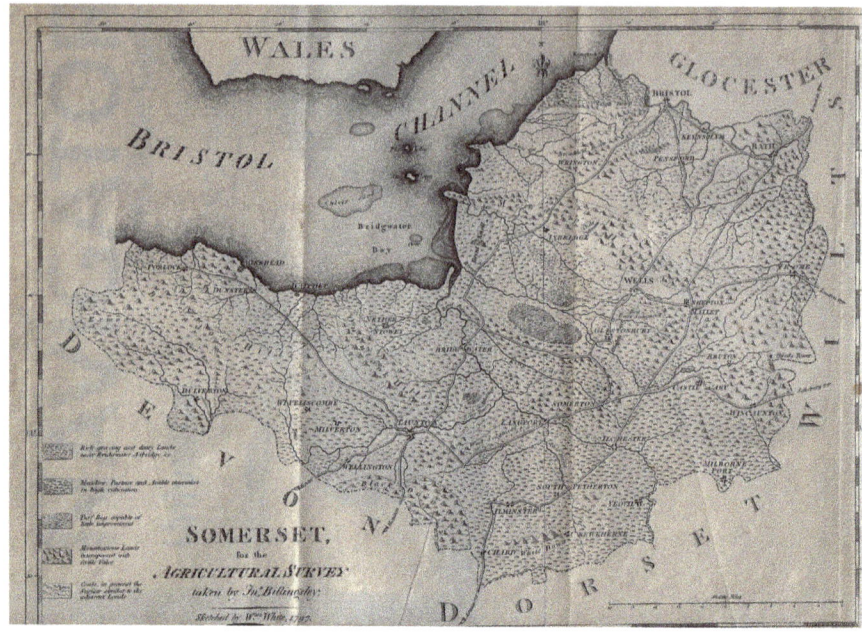

Map drawn up by Wm White, for Billingsley's General View... second edition, 1794[10]

this method would presumably have been acceptable, but this approach was not helpful as it was part of a series for which the board explicitly requested that the information should be sequenced so as to enable readers to easily find and compare sections in different county reports. The objection that the whole is uneven in its coverage is indefensible. As a positive, the second edition contains many updatings, particularly in prices – helpful as this was an era of Inflation.

As previously remarked, Billingsley often gives no headings to help signpost the reader, or gives signposts that are somewhat misleading. Towards the end of the report he gives 'A Recapitulation of the Hints for Improvement already suggested in the preceding pages'. This is the most obviously useful part of the survey, discussed below. The length of Billingsley's first edition was about double the average number of pages for all reports in the first phase. This can be accounted for partially by the fact that he did address, at least to some extent, all the headings the board required (which not all reports did), albeit his were not necessarily in the order or place requested. For many items, though, his report is admirably detailed. It seems that the board thought Billingsley's *General View..* was one of the best received.

However, Billingsley had also added additional information not requested on some topics: for example, he set out in full the 'Laws and Orders of the Mendip Miners', dating from the time of Edward IV (mid-fifteenth century), which while very interesting and at the time still adhered to in principle, does not seem to be essential for agricultural knowledge or practice. Noticeably, though, he did not pad out the text by, for example, detailing the lands of the nobility, as did some authors (such as Vancouver in the *General View of Devonshire*). Billingsley's report is essentially practical. Incorporating extra material meant that the second edition reached well over 300 pages. It initially sold for six shillings, the price set for the surveys intended by the board to only just cover the costs of printing and distribution so that as many people as possible could benefit from it. This made the whole project very expensive for the board, which soon experienced financial difficulties.

As noted above, the most important part of the report was the final section. This is especially true for the information on practical agriculture. In both editions Billingsley replaces the board's heading 'Obstacles to Improvement' with 'Hints for Improvement' – a change unremarked, and arguably an improvement in itself (although the lack of any discussion on obstacles to improvement is something of a loss, especially if using the report to compare this county with others, as had been part of the intention). The 'Hints for Improvement', are, as he says, a recapitulation of his advice at various points in the book, drawn from the aforegoing text, which has mainly described things as they were at that time. They represent Billingsley's knowledge and skill in improving agriculture, and carry a large amount of the best of the advice from the report, so are worth looking at in some detail. In most cases there is little variation between the content of the two editions: if no comment is made here then to all intents and purposes they are the same.

'1st: Inclose and cultivate all waste lands susceptible of improvement, and divide and inclose the common fields'. In this case, apart from the heading, the two editions do differ quite significantly. In the first edition he supplements this 'Hint' with the claim that, whereas the lands of the Mendip Hills were originally worth £1,500 in total, following enclosure they are now worth between £7 and £8,000 per annum. Yet there are complaints, he says, that the price of corn has not diminished – this is due, he thinks, to its scarcity. The poor rates have not diminished, partly due to the increase in population but also due to what Billingsley sees as the indigence and profligacy of the poor, following this with one of his now familiar (and lengthy) diatribes against the labouring class. He advocates 'Friendly Societies' as a means of ensuring families survive distressing periods. While the second edition has the same first 'Hint', his section on the poor is gone, replaced by the potentially more useful suggestion that landowners should 'apprentice' their sons to learn farming in the way that the young are apprenticed to learn a trade. He may have decided against including his view on the poor, acting on negative feedback, or the board requested that it should be omitted.

'2nd: Where lands are situate on bleak and exposed eminences, improve the climate by judicious and extensive plantations'. In both edition's Billingsley states that in these situations 'massy plantations' were not only ornamental but profitable. Scotch fir was probably the best, but also sycamore, ash and birch; to be planted on the south-west of a farm (the windward side). But they must be protected from stock.

'3rd: Wherever marl, chalk or lime [are plentiful] neglect not to make a liberal use thereof; and if destitute thereof, be careful to make as much dung as possible by folding sheep, housing…cattle, and preserving and collecting' all sorts of animal and vegetable matter to add to the soil. Although Billingsley does not actually use the words 'compost' or 'fertiliser' he explains in some detail how to manage the making and application of what we would now recognise as these substances, saying that 'by these means a strong fermentation is excited'. In the northern part of the county marl, chalk or lime are easily obtainable, and their application need not be repeated for several years, but such additions are sometimes unused, and there are places where these items cannot be found. There more dung and compost must be used. He shows a good understanding of the need to fertilise soil, even though at the time the chemistry behind it was unknown. In the second edition he adds that 'in the application of dung, farmers in Somerset go about it from the wrong end', applying manure to the land before a wheat crop… 'whereby the wheat crop is made foul, and though there is great burden of straw there is but little corn'. It would be better to apply it to the potatoes and turnips, he says, and make wheat the last crop in the rotation. Also, applying it so that it overwinters on the land is helpful, which has been generally recognized by others only relatively recently.

'4th: A Universal and regular rotation of crops'. As in many other places in his writings Billingsley bemoans the 'improper conduct…of the tenant. No sooner is the plough put into his hand, than he uses it without mercy.' This leads to landlords counteracting progress with restraining orders so that ploughing cannot take place even in cases where it might improve the land. Some old grassland is mossy, hidebound and unproductive. In this case the plough would be used with benefit. The preferred rotation being:

On light ground:
1st peas or oats on the ley
2nd vetches fed off, and the land manured with lime or folded sheep

3rd barley and artificial grass seeds
In which state let it remain till the grasses fail, and the land again becomes mossy; then renew the course.

On heavy ground:
1st beans on the ley
2nd spring fallow, well manured; and cabbages
3rd oats and artificial grasses
Then remain as before.

To which he adds that on heavy land two tons of cabbages are equal to three of turnips, which suffer no injury from frost and the expenses of cultivation are less. It is interesting that he advocates a four-year rotation here – while in his 'Essay on Waste Lands' (see chapter 15) he suggests a fourteen-year rotation for reclamation of waste land.

'5th: Enlarge the upland corn farms; erect proper buildings and conveniences for the shelter of the cattle in the winter months, thereby inviting substantial, and well-informed farmers, of more enlightened countries, to settle upon them'. Nothing so much contributes to the progress of good husbandry as large corn farms, buildings and so on. And 'the best method by which general improvement can be promoted is by dispersing the farmers whose practices are held in the highest estimation among those parts of the kingdom, on which the light of good husbandry has never shone'. This would introduce the turnip grown in the eastern district of the country to this south-western area, he says, where the soil and climate are so well adapted to its growth.

'6th: Improve the stock by a judicious selection of animals for breeding; and be particularly careful to choose a male handsome in those points wherein the female may be deficient.' Although the farmers of Somerset are aware of the importance of good quality stock they are 'very inattentive' to it, he claims, so that their breeding males leave much to be desired. If bringing in stock for improvement then they should not be moved from rich to poor land, or from a warm to a cold climate.

'7th: Encourage the use of oxen', he says in the first edition. But by the time of the second he strengthens this to 'lessen the number of horses'. He was no great lover of the horse for agricultural work, claiming that 'it is universally acknowledged that too great a portion of land is employed in raising food for horses'. He had strong opinions on the superiority of oxen over horses for ploughing and other uses on the farm, plus they eat less and once over their working lives the carcasses of oxen are worth more than those of horses. He expresses this opinion in many places in his writings, although it would appear his battle with the agricultural community on this front was already lost. Even so he goes so far here as to say 'every man who keeps an unnecessary horse is an enemy to his country, by reducing the increase of his own species'. He then remarks that navigable canals will reduce the need for horses as pack animals, and canals are beneficial to the public.

'8th: Amend the public roads'. This is another of Billingsley's oft repeated mantras: 'nothing so much contributes to the improvement of a county as good roads', a view based on uncomfortable personal experience, in more senses than one, as 'before the establishment of turnpikes, many parts of this county were scarcely accessible'. He describes the way more horses had been needed to haul goods, especially heavy goods like coal, as in his own area. He also again criticises the system for maintaining 'private' (non-turnpiked) roads, advocating a highway tax for their upkeep. In the second edition he also mentions 'road clubs' which, he says, have been beneficially established in some places, but he does not describe their operation.

'9th: Encourage the use of such ploughs, and other instruments, as are best calculated to expedite work and do it well'. Another repeat of one of Billingsley's cherished beliefs: the superiority of the double-furrow plough over the common plough or wheeled plough. He describes how in Somerset there would be great savings from the use of the first of these choices. However, he apparently did not manage to convert many Somerset farmers in this instance either.

'10th: Sow early in exposed, and cold situations, and be particularly careful not to plough or harrow in wet weather'. This rule is so well known, he says, that he does not need to enforce it.

'11th: Destroy rats and mice'. The depredations of these vermin are too important to be overlooked. In the second edition he puts a figure on it: 'between these vermin and birds a thirtieth portion of corn in the kingdom are devoured.... It behoves all farmers to make their slaughter a general concern'.

'12th: Introduce threshing machines'. These are used in northern areas of the kingdom, he says; the only objection to them being the lessening of indoor labour in the winter. Housing all cattle in winter would mean this labour could be used in attendance on them. Billingsley championed machinery in all cases.

'13th: Let all unmalted corn be sold by weight'. Complications of different measures for corn have led to the Winchester being in general use, which in turn had led to the benefit of the seller and loss of the purchaser.* He

---

* The Winchester bushel was the traditional corn measure, a legal standard of volume introduced in 1495.

believes it would be fairer to sell corn by weight rather than quantity, as the drier and plumper corn is the heavier it weighs.

'14th: Grant long leases'. He says that 'all farmers who have spirit enough to improve their estates should have some security.' If a man's tenure is precarious then little improvement can be expected. If a farm is unimproved then there will be considerable expenditure necessary to improve it. In this case the lease should be 21 years, with increase in rent at 7 years and 14 years. He suggests that the landlord might pay all the expenses of improvement, charging 5% on the expenditure, allowing the increase in rent to be proportionally less. But this is 'so copious a subject that I must forebear entering into it' – it would be ably treated by numerous correspondents, he suggests.

The first edition follows this point with the 'Conclusion' – but the second has a further five 'Hints', related to cultivation of specific crops (one of those places where, towards the end of a paper Billingsley seems to become aware of all the things he had forgotten and adds them hastily). These are: 15th: sow more sainfoin; 16th: roll all grassland at least once a year; 17th: set peas and beans in lines north and south and hoe them twice (seed was still usually broadcast at this time and hoeing was in Billingsley's view sadly neglected); 18th: devote at least one quarter of turnip land to ruta-baga or Swedish turnip (now known as swede, which he sees as superior in performance to common turnips).

His final point is that this county appears by every measure better adapted to grass than arable, and enquiry should be made as to whether animals cannot be kept here on grassland alone, without the aid of winter roots. The plough is usually used to produce straw, he claims, but in the hands of the generality of farmers the land will soon be in a state of degradation. 'Grass, therefore, should be seen as the ultimate improvement of land in the Western part of the county of Somerset'. Presumably he feels that other parts of Somerset will benefit from some land remaining as arable.

In his 'Conclusion' Billingsley states that 'this county does not raise grain sufficient for its consumption, nor are the climate and soil of many parts thereof favourable to corn farming', although he believes that if all the improvements mentioned were undertaken the produce of the soil might be increased by one third. Chief among these improvements are, of course, Enclosure and drainage. In both editions he puts a figure on the increase in rent achieved by these improvements: from the drainage of 115,000 acres this would amount to £43,750 per annum (increased a little from the first edition); from Enclosure of land possible for arable or pasture: an increase in rent of £213,000. Finally, he uses the same 'Postscript' for both editions, making an apology for his tardiness and an admission that he had not expressly undertaken a survey for this work, due to his prior knowledge of the county. He dates the postscript 4 October, 1794 in both editions. His second edition was again well received by the board.

Following the publication of the majority of the county second editions the board evidently requested that William Marshall undertake a *Review and Abstract of the County Reports of the Board of Agriculture*.[11] At first sight this is a little unexpected, in view of Marshall's treatment by the board in putting Young in place as secretary rather than him. Perhaps the fact that Young had ended up writing several of the county reports himself influenced that choice. More likely it was the result of Young having recently lost favour with the board.*

Marshall's *Review* covers all county reports, running to five volumes. The great majority of it comprises quotations from the various reports of sections he thinks useful or remarkable. Occasionally he signals agreement, but more often where he makes comment it is not positive. He is generally extremely critical of reports and reporters alike, perhaps a natural reaction in view of the slight he had suffered from the board (his views on Young's contributions are particularly scathing). However, Marshall's own experience of agriculture in the south-west was limited, so his opinions are not necessarily reliable. His most important – and perfectly valid – criticism of the surveys is that they covered counties, rather than farming regions. This certainly reduces the usefulness of the series.

In his assessment of Billingsley's report, and of his capabilities overall, Marshall is grudgingly positive. 'Mr B... is a man of considerable ability and information.... as a Reporter of rural concerns... Mr B's experience in agriculture has arisen (in one particular at least), from an extensive practice.' He recognises Billingsley as a 'drainer and encloser' but he is either unaware or chooses to be silent on his reputation as a practical agriculturalist, or of his many well-received papers in the *Letters and Papers* of the Bath and West Society, and other publications. 'It will be apparent to professional occupiers,' he says, 'that Mr B's strictures are those of an *improver*, rather than one who has long been in the habit of paying *personal* and close attention to the *minutiae* of practice'. This judgement was not based on fact: Billingsley's expertise was

---

* Having initially been a champion of Enclosure due to its agricultural effects, Young changed sides following his recognition of its effect on the poor: full employment and adequate wages did not result from Enclosure, he now believed. This ran counter to the board's position, and was the precursor to them marginalising him from 1801, when they refused to publish his report on a 'Tour' which had been undertaken at their behest.

definitely one of the personal and closely attentive variety, which had led naturally to his equally valid expertise as an improver. Marshall goes on to make the general point that 'a mind accustomed to consider the outline and general economy of an art is more likely to afford a comprehensive view of it, than a mere practitioner professor – who, on the other hand, is better qualified to describe existing practices, so as to convey them'. Marshall notes that 'Mr B has obligingly favoured his readers (contrary to the usage of most others of the board's reporters) with some information concerning the work he was offering'.

Young, on the other hand, was very aware of Billingsley's practical experience: he (and his work) was personally known to him. At appropriate times Young gave a more rounded – and positive – critique of Billingsley's capabilities, for example in his various letters to the society. Moving forward to 1850, in that year the new Royal Board of Agriculture of England (established after the previous board had been discontinued) offered a prize for an 'Essay on the Farming of Somerset'. This was won by Sir Thomas Dyke Acland.*[12] In this essay, and elsewhere, Acland is very complimentary to Billingsley, specifically mentioning the latter's 'Hints for improvement' in his *General View...* claiming that 'Mr B concluded his report with certain practical suggestions...the best farmers will bear me out' in that there is not one which does not apply 'in all its force to some part of the county at the present day'.

Billingsley's reputation today rests in large part on the *General View...* Unfortunately, though, those who have written on its agricultural aspects recently have either tended to repeat Marshall's rather limited assessment of him as a 'drainer and encloser', or more commonly they have pointed out his undeniable prejudice against the poor, dismissing his agricultural usefulness.[13] To those then engaged in agriculture, it was undoubtedly of very great use.

It is impossible to resist quoting Billingsley's final sentence in the *General View..* as it has such resonance for today's world, agricultural and otherwise: 'Would to God that nations would learn wisdom, and instead of coveting distant territory, improve to the utmost that which they possess!'[14]

---

* Sir Thomas Dyke Acland, 11th Baronet (1809-1898), was from a prominent family with extensive estates in Devon and Somerset. Among other things he was a politician; interested in agriculture; trustee of the Royal Society of Agriculture, wrote the prizewinning essay 'On the Farming of Somerset' (1850); and is credited with reviving the Bath and West Society.

15
AN ESSAY ON THE CULTIVATION OF WASTE LANDS

SOME YEARS AFTER the publication of the *General View..*, Billingsley's 'Essay on the Cultivation of Waste Lands' (called 'Waste Lands' hereafter) won the Bath and West of England Society's prestigious 'Bedfordian Medal' of 1806 and was published in volume XI of their *Letters and Papers* the following year.[1]

As argued earlier (chapter 7), in many ways this essay may be considered to be almost as significant as the *General View..*, though it is certainly not as well known. It has been paid very little attention since the mid-eighteenth century, perhaps because it was not part of a nationwide series. Yet as the last of his many useful papers on farming methods it is definitely the best. The *General View..* covered the whole of Somerset and a broad scope of agricultural topics so it is inevitable that Billingsley will not have had personal experience of some aspects (although in this respect he was far superior to most of the other survey authors). By contrast, in 'Waste Lands' he wrote about matters of which he had deep personal knowledge and experience, most of the content being, in effect, a practical handbook for bringing Enclosed land into cultivation – and keeping it productive thereafter. As remarked previously (in chapter 7), while every county had a 'General View', Somerset is probably the only county to have a handbook of this quality on how to achieve significant advance in the agriculture of the time.

As so much of his expertise came from work on the exposed lands of Mendip his advice might be seen by some as applicable only to areas with similar conditions. However, if sensibly adapted to meet conditions elsewhere the gist of his advice would generally still hold – at the very least in the south-west. His methods are useful for reclaiming almost any previously unmanaged land. Thus, 'Waste Lands' would be of more use to those who might be persuaded to Enclose, or seek to improve newly Enclosed land – his intended audience – than would the *General View...* The actual process of Enclosure is given prominence near the beginning of 'Waste Lands', but the act of improving land, the subject of the later part of the essay, is potentially relevant to all land. The text is also rather better organized than most of his writing, making it easier to glean the crucial principles. Billingsley's ideas and practice as described here were ahead of others of his time.

There have been well over 200 years of advance in knowledge and change in practice since the essay was written so, in addition to its perceived merit at that time, another measure of its value is the extent to which the essence of his ideas, methods and their results is still relevant today – especially to those most engaged in agriculture. As published, 'Waste Lands' runs to over 90 pages; a synopsis is given below, with discussion of each section. Some comments from other writers are added here, ranging from that time until the present. As the essay was aimed at those who might be encouraged to improve their land, those active in improving agriculture today are the most appropriate to evaluate it: local farmers and agriculturalists (some of whom farm on land once owned by Billingsley). Their views given here are mostly confined to the second, more important part of the essay. This chapter therefore gives a different perspective on Billingsley's work in agriculture to other parts of this book. Farming is obviously very different today, the main changes coming from advance in understanding and application of science to agriculture and the introduction of machinery, yet many aspects remain pertinent.

The essay begins with Billingsley complimenting agricultural societies, the Bath and West Society in particular, for their role in Enclosure, and stressing the urgent need to do more. He has economic need clearly in mind:

> a general system for facilitating the division and improvement of waste land and commons is an object anxiously to be wished for, not only as a means of promoting general prosperity and also increasing the growth of grain, and thereby preventing a return of those years of scarcity… which had they occurred at the present time (when almost all the ports are shut against Great Britain) would have been fatal in the extreme.

In this time of war thought must be given to the country's ability to feed itself.* But, he says, there is still much opposition to Enclosure and little hope of help from parliament. Some way must be found to convince the 'cautious investor' that this path will be profitable to himself as well as to the

---

* Today, the ability of our country to feed itself is again a prominent issue, though for other reasons. As then there are consequences for farming.

community. By the year in which Billingsley is writing most of the Enclosure acts for Mendip had already been passed or were in hand, so he probably had other areas in mind – such as Devon, which he criticises elsewhere for falling behind in progress on Enclosures.

He continues by addressing the objections given to enclosing 'waste', namely:

> 1st the heavy charges of allotting, enclosing, cultivating and manuring etc will be insufficiently repaid by the increased value
> 2nd the long held opinion that new enclosures will be profitable for a few years, then will "return to their original sterility". Even those who strenuously advocate enclosure in lowlands are violently opposed to cultivating the hills
> 3rd upland commons are seen as nurseries for stock without much expense and ought to remain so
> 4th enclosing wastes has the tendency to depopulate the country and deprive the cottager of those privileges and advantages which are his birthright and without which he cannot comfortably exist
> 5th some landowners have claimed that an increase in produce, especially of oats, would glut the market and reduce prices, therefore lower the rent of old estates. But the experience of the last 30 years, during which the price of oats has progressively risen "negates this objection".

Billingsley claims his essay will refute the 1st and 2nd objections. As for the 3rd, he says, those who use commons for pasture must have noticed they are so overstocked that stock does not thrive, and wild barren commons will not provide the same advantages as Enclosures. Overstocking of commons was claimed by so many observers at the time that it was obviously a very real problem.

Billingsley approaches the 4th objection in characteristic fashion. He was not one to avoid contentious issues – rather he revelled in making his views public. This part of the essay is admittedly contentious. The situation for the poor was an issue which polarized opinion at the time: it was a period notorious for its harsh treatment of the poor, even though reformers were active. He claims that while the 4th objection 'implies such humane concern for the rights and comforts of the poor' that it deserves greater and more serious consideration, he considers that the fears expressed in it are 'not justified by experience'.

How can it be possible that residents of a district can be lessened in number, or the poor injured by a measure which secures to the peasant advantageous employment and enables the farmer to pay better wages than before?

Though the population certainly did fall in many Enclosed areas nationally, it is also true that the population did not fall (or fell only marginally) in most of the land Billingsley was concerned with. In any case most of these areas previously had a very low population density. The population on Mendip, which had been unusually low, actually *rose* as a consequence of Enclosure – though admittedly this was an uncommon circumstance.² Billingsley gives no evidence of better wages being paid after Enclosure; in fact he does not address wages or employment at all in this paper; nor does he really address commoners' loss of rights. He is decidedly dismissive of their difficulties. His view is that following Enclosure:

> in all instances (and they have not been few), the cottager has gradually risen from a state of sloth and poverty to active labour and comfortable subsistence. His house, which was before the miserable hovel of disease, wickedness and want, has assumed a clean and comfortable aspect, and his children, who were often without any employment and addicted to stealing and pilfering, have been furnished with the opportunity of contracting habits of industry.

Billingsley seems to be genuinely convinced of the benefits of Enclosure for the poor, although he sees things from the point of view of the agriculturalist and has very little empathy for the commoner or the poor in general. However, there are signs that at this stage in his life his conscience may have been pricking him, for example, rather more attention to the issue of poverty than in most of his earlier writing (and inclusion at the end of the second edition of the *General View.*. of a late section on the question of provision for the poor in Enclosures). It should be said that many other members of the Bath and West Society (including its first two secretaries, incidentally both quakers) were very aware of the plight of the poor at that time and sought to have the society become more active in that respect – Billingsley notably not among them.³ Poverty was a divisive issue at the time in both the society and nationally. Many with his outlook felt that landless country people would always be in a state of on-going poverty, through their own laziness and profligacy, regardless of any help given them.

He goes on to say that: 'in most cases a considerable amount of garden-ground has been annexed to [the cottager's] dwelling, and in some instances land sufficient to keep a cow or two has been allotted him.' There is evidence of some provision of land for the dispossessed, and a few labourers' cottages were built on land he Enclosed, as he mentions later in the essay, but this was not done in all cases. The Bath and West Society offered premiums for the design of such cottages.[4] Land was also set aside in some parishes for the benefit of the poor through the poor law overseers or vestry, but not in all. Billingsley certainly built 'comfortable farmhouses', but these were for farmers of sufficient substance to rent the land, not for cottagers. He recommends hiring by the job rather than by the day, 'for it must be mortifying for the active labourer to see the same daily wages paid to an awkward and lazy vagabond as to himself", thereby promoting insecurity of employment for labourers and greater cost effectiveness for employers. Overall, while he has not totally shirked the subject of the social effects of Enclosure, his defence is wanting: his attention is fixed on the agricultural benefits.

Billingsley goes on to say that he presumes the society is offering a premium on this subject in order to encourage landowners in the west country to undertake Enclosure, especially as so much waste land in the region is in cold and bleak areas, not conducive to easy farming. He stresses his experience, having Enclosed several thousand acres, in a climate 'unfavourable to corn husbandry', his first attempts 'attracting ridicule' from some, while his friends applauded his courage and perseverance but 'predicted his ruin'. He tends, he says, to underrate the average quantity and value of the produce that results, in order to take into account bad seasons and other accidents. He is convinced of great profit as well as satisfaction in seeing a barren waste converted into a fertile Enclosure, and healthy employment for families on a spot where previously there were scarcely any inhabitants.

He then moves on to the crucial part of the essay, with a subject where he is at his best – a very detailed description of how to go about the actual process of Enclosing with a view to optimising the state of the land. From this point onwards almost all 'Waste Lands' concerns the practicalities of Enclosure and subsequent work to improve the land, together with the costs and financial rewards. (As far as costs are concerned, it is notoriously difficult to compare costs then and now, therefore no attempt is made here to decide whether the figures he gives are likely to be high or low, though some comparisons can be made within the costs shown.) It will be obvious to most, but even so it is worth noting, that even within a small area the land can differ significantly in farming terms. For example, taking Mendip, as an overgeneralisation the 'plateau' can be seen as one region, while the ring of land around it, still part of Mendip, forms another. And even within each of these regions, the type of soil and other conditions can differ substantially from one field to another.

Billingsley held and cultivated land in both Mendip areas, and was very familiar with the Levels, different again, his experience being drawn from a wide variety of farming situations. The advice he gave would be relevant to all these different situations.

Billingsley divides the subject into five parts:

1 Enclosing and Dividing This includes outside boundaries and interior 'fences'. He details types of 'fence' (meaning boundaries): stone walls; hedges; banks and ditches; dead hedges; how to build and maintain the various types; costs of each (mortar walls 17s 6d; list walls [stone walls with turf topping] 15s; dry walls and quick [hawthorn] 9s 6d; banks and quick 11s – each per rope of 20 feet); types of plant suitable for hedges; all with pro's and con's; use of trees in enclosures. Billingsley regarded the stone walls as temporary while hedges were being established. He understood (as many still do not) that hedges are superior to walls for stock shelter. Yet a large proportion of these walls remain on Mendip today; they give a bleak appearance, but farmers have not replaced them. Today, after many years of grubbing up hedges (grants were available for the purpose), they are now greatly valued for their ecological advantages – though this aspect was scarcely thought of in Billingsley's time. Some hedges are now being replaced (grants now available for the purpose). He stresses the need to keep stock away from young hedges, then lists types of plant for hedging, but making no mention of hazel, the predominant Mendip hedge tree today, ideal for laying. Many of these methods are still in use, for example ways of building stone walls and laying hedges. The method he gives for moving semi-mature trees is also back in use today, having been re-popularised in the 1960s (but still seldom used by farmers locally). In some cases his advice was controversial at the time, such as the need to plant trees at intervals along the fence to provide shelter for stock and for a future supply of wood. In addition to their use along fences Billingsley usually planted shelter belts of trees around Enclosure farms, as at Green Ore. These belts are still common across Mendip today, one of the most apparent visual features. Trees can be used, he says, where the ground is unsuitable for other purposes. His list of suitable trees omits elm, which until the ravages of Dutch elm disease was probably the most common Mendip tree.

He recommends field sizes of 20, 30 or even 40 acres for land that will be arable, regular shaped and usually rectangular for ease of cultivation. Michael Williams points out that immediately post-Enclosure field sizes on the Mendip plateau proved to be too large so that many were subsequently subdivided again.[5] This was often where the plots were used for potato growing, leased to the landless poor. Post-Enclosure fields contrast with those that existed in the district before Enclosure, most of which were small and irregular, often as little as two to five acres. The difference in shape and size between pre- and post-Enclosure fields is obvious from a map, or from walking the land. Average field sizes have now increased again, for ease of use with modern machinery. But, Billingsley says, before internal divisions are made fields should be prepared first by ploughing and manuring, and drainage put in using ditches where necessary.

2 Buildings 'The principal object in farm buildings is convenience'. He lists the types of building necessary for a farm, where stone is easily available (as on Mendip). The house should face east or south-east, the roof be made of oak with tiles or slate covering, the optimum size 42 feet in length, 18 feet width, and 22 feet in height. As an exemplar he provides a description of Lord Waldegrave's farm at Chewton Mendip (Wigmore Farm). Farmyard buildings should include: stables (which should be furthest from the house in case of fire); a barn (if possible on sloping ground); open stalling for cattle, cows and calves; a waggon house and granary; pig styes and poultry house; room for a threshing machine if wanted (this was a new invention at the time). The actual costs of the buildings at Waldegrave's farm (which Billingsley was responsible for building) were: farmhouse £499, stable and hayloft £110, barn £160 and other buildings, all adding up to no more than £1,000. It is noticeable how many of the farmhouses and associated structures put up on Mendip during the Enclosure period have persisted relatively unchanged (from the outside at least), despite the introduction of modern methods and machinery, though extra buildings and barns have usually been added since for further and larger storage. The original stone buildings were long-lasting and well suited to the environment – if damp and uncomfortable by modern standards.

3 Cultivation and Manuring The first task, Billingsley says, is to move rocks and stones from the fields – still a significant task even with today's machinery. In many places the results of mining (mineral 'batches' or rubbish piles, pits and hummocks) all needed to be attended to. A local archaeologist recently remarked that Billingsley can be credited with 'making the landscape of Mendip beautiful again' after the ravages of ancient mining (though there are still some signs of mining even today).[6] Furze (gorse), ling and heath may be burnt or taken back to the farm for fuel. Once the surface is clear the land should be ploughed, cross-ploughed, dragged and harrowed. Next, he says, it should be limed: he gives details of how to go about this in various circumstances; quantities and types. He recommends slaked lime, presumably because it acts more quickly than unslaked. Mendip land is very hungry for lime as despite being on limestone bedrock the high rainfall dissolves any added lime which percolates down through the soil. Lime should be used as a soil conditioner he says, making the soil less acidic thus enabling nutrients to be taken up. It is still used today. Acland claimed (in the 1850s) that by adding sufficient lime only once the effects would last fifteen or twenty years.[7]

He gives details of the total expense per acre from the first ploughing, including dragging, harrowing, liming and more – through to a fourth ploughing – costs amounting to £6 12s an acre. In this state it may be let to a tenant or used by the Encloser, he says. By any measure this is thorough preparation. Many farmers today consider it too thorough, ploughing being destructive to soil life. In some cases ploughing in itself can be problematic: not just stones in the soil, but the bedrock often being so close to the surface: 'Mendip comes up to meet you' is the phrase used. In these places, despite the high rainfall, the land can drain too fast for corn, one factor in the demise of cropping wheat in particular. The old wheat types would have been more affected by wet and windy weather than modern varieties, as they were very susceptible to mould and disease and more likely to be lodged (thrown down) than current varieties. Another difficulty on Mendip in particular is that water for stock would have been a major problem even with Enclosure, as without it land was only fertile near watering places, seriously limiting stocking densities. Availability of water is still a local problem for Mendip. But Billingsley does not mention any of this as an issue; he was positive by nature.

As 'practical results are more convincing than imaginary statements', he returns again to Wigmore Farm in Chewton Mendip with an example of actual costs and profits. On this farm of 440 acres, 50 of those acres were overly encumbered with mineral batches, so unploughable, meaning only 390 acres would give best return. A detailed table of all costs is given, leading to a statement of profit of £4,200. The estate was then let on a fourteen-year lease. This was a time of inflation: costs have risen since it was let, he says. He estimates that in the four years since the farm was built costs would then total £100 more.

4 Cropping and Harvesting Some people, Billingsley says, claim the statement of profit above is fallacious. He regards the problem here as being the common practice of 'six or seven successive corn crops without intervening fallow or variation, taking the produce home': while the farm temporarily does well the Enclosure is 'beggared'. The failure to heed this advice, which at the time was endorsed by many others, was a major issue with Enclosure land. As late as 1850 Acland says that farmers on newly Enclosed land 'grew oats without manure as long as the land would bear it'.[7]

The 1801 crop returns give some information as to land use in

Billingsley's time.[8] The new Enclosures certainly grew an unexpectedly large amount of corn: in East Harptree, for example (where Billingsley held land), corn accounted for more than 88% of land in crop that year, with over 65% of this being oats (which are particularly well suited to the area's terrain and climate). In some parishes nearby the figures were even higher. Unfortunately, we do not have figures for other years. Obviously no single year can be taken as showing a trend, but if this were repeated it does seem that the land could soon have become exhausted. To secure against this fate Billingsley suggests a fourteen-year rotation (shown below), at the end of which the land will be totally productive rather than exhausted. The idea of a fourteen-year rotation produces an almost unanimous response from today's farmers: fourteen years seem excessive! Not for it not being productive, rather because it would be unnecessarily complicated, requiring various types of machinery and management, plus different markets for the produce. This would not have been so much of a difficulty in Billingsley's time.

Having recorded detailed costs for his preferred rotation Billingsley then demonstrates (with figures) the effects of a seven-year system using corn crops more frequently – the resulting profits are £12 for the fourteen-year rotation and only 6s for the seven-year; this calculation 'which I pledge myself to support by incontrovertible facts... [should] be sent to tenants who still practice this abominable system of scourging their lands'. It must be a rule not to have more than two corn crops in succession, he believes, for doing so would mean the land is 'rendered foul and unproductive'. The 'evil effect' of overcropping leads to the quantity of corn land being greatly diminished and this has a double effect on the labouring classes as 'it increases the price of their principal article of subsistence while it circumscribes their means of purchasing it'. There were many occasions when this advice was ignored, with serious consequences. The land then needed reclamation all over again.

ROTATION of CROPS.

1st. Wheat.
2d. Potatoes, turnips, or cabbages, (dunged lightly.)
3d. Barley, or oats, with clover-feed.
4th. Clover, (broad or red clover.)
5th. Wheat on one ploughing, and after two mown crops of clover.
6th. Vetches and turnips, (well dunged.)
7th. Oats, barley, or buck-wheat.
8th. Ray-grafs, and white Dutch, (fed.)
9th. Ditto, ditto, (mown.)
10th. Ditto, fed, and manured with lime and earth.
11th. Ditto, (mown.)
12th. Oats on one ploughing of the ley.
13th. Vetches and turnips, well dunged; or peafe.
14th. Barley, or white oats, and feeds.

Billingsley's Fourteen-Year Rotation of Crops for Land Claimed under Enclosures, from An Essay on Waste Lands [9]

Billingsley's period was the start of the serious use of informed crop rotation, with him at the forefront of the movement. Nowadays rotations of between two and four years are more often used on arable land – but that of course is with modern understanding of soil and fertiliser requirements, plus farmers today have the benefit of the land being already improved. There was recently a trend towards monoculture continuous cropping but that has now been returned to a more sustainable system, with three crops in two years.

Billingsley believes that in convertible tillage (in which land is converted from pasture to arable and returned to pasture) the land will be more productive than under permanent grass, the profits greater and the farm progressively improved. Conversion is much less often practiced on Mendip today, most grassland being permanent pasture. In fact, relatively little arable remains on Mendip today compared to immediately post-Enclosure. Regardless of its preparation the land there, as Billingsley predicted, is better suited to grass.

A detailed account of expenditure and receipts during reclamation is then given, so that one can compare the expense of the produce of an acre of land: in the first seven years of his recommended fourteen-year rotation the profit is nearly 35s per acre; the land becomes worth 10s 6d more per acre than originally. Whereas, using seven years of grain the first two years show profit, but thereafter yearly losses – with a balance of 6s profit only after seven years. A main reason given for taking successive corn crops is insufficient capital, Billingsley claims. A man with £500 will be keen to take on a farm at £200 rent but after purchase of the necessary implements, stock, seeds and so on, he will then be in a bad way financially and 'will set the plough to work without mercy', so will need to send produce to market immediately whatever the price. Every practical farmer knows that on all poor upland farms, grass (if the soil be favourable) is more profitable than corn, because the expense of cropping arable land is certain but the return uncertain... and the fewer acres a farmer has in tillage, the more he may raise... ten acres under a masterly system will produce more grain than twenty acres slovenly managed.

Billingsley adds that he does not wish to exclude the plough, but maintaining the proper proportion of arable to pasture is both useful and necessary. There should be no more than one third part in tillage. This tenet is repeated time and again in his publications.

5 Succeeding Management Managing grassland is simpler than arable, he says. 'Artificial [non-native] grass generally thrives on newly cultivated land manured with lime, though broad (red) clover takes the lead, ray-grass and white Dutch clover are best for permanent pasture.' Again, clover is still regarded as excellent today; and the scourge of couchgrass he mentions is still present, though blackgrass is now even more of a problem and difficult to control. Today's improved crops stand up to the weather better than those used in Billingsley's day. In the eighteenth century they did not have the benefit of recently introduced strains of ryegrass; today these are sown in autumn and mown three or four times in the following year. Native grasses are now favoured by many farmers on Mendip, for ecological reasons. It is likely, though, that Billingsley would have championed introduced varieties as they are more profitable, and, like others in his day, he had very limited awareness of ecological issues as we now understand them.

Moving on to sheep farming (one of the few places where he discusses stock), Billingsley describes the local system of lambing in February at lower levels then sending the sheep back to the tops in late spring. Also, 'Scotch runts' can then be purchased and kept for the winter months. If they are fed hay, vetches and turnips then there will be grass available on the high ground in summer. Though he does not mention it, here is an example of the way a small farmer of his time would be disadvantaged without common land, as he is unlikely to hold land in both places.

The 'prudent farmer' should grow no more corn than is necessary to renovate his old pastures and supply the farmyard with straw. He says that the straw is the intended crop: to the non-farmer this is rather unexpected, especially as it was a time of high prices for grain. The farmer should mow the grassland as little as possible and keep no winter stock on his newly sown grasses. Billingsley then gives instructions as to the best methods to feed stock with the greens grown during the rotation. Barley straw, with the softest stalk of the cereals, can be fed to stock in the winter (this was a time before silage, now used for winter feed); other straw can be used for bedding. Returning to the annual work programme he considers necessary for success he admits that 'this masterly plan will, no doubt, be considered as unnecessarily profuse... the expenditure included... but no other plan can ensure permanent success'. He suggests some alternative crops, such as potatoes and sainfoin, and how to manage consumption of crops by stock. Alternatives to grass are often used today, especially maize and turnips, to supplement grass crops (fodder beet is less used on Mendip recently as it has harvesting problems in wet autumns).

Billingsley's last section is entitled 'Miscellaneous Observations'. Establishing a nursery to raise trees and setts (for hedges) should be the farmer's first priority, he says, with a sufficient sum of money set aside to enable him to proceed 'without restraint or embarrassment'. He must give 'unremitting personal attendance' to his land. He recalls that William Marshall has observed: 'Farmer Self-attendance is the best farmer'.[10] The gentleman farmer should proceed by contract and hire an honest and capable bailiff and a good shepherd. The time of sowing must vary according to the season, but wheat must be sown by the middle of October in order to produce a sufficiently early crop. Attention to the purchase of good quality seed and keeping it in good condition is also stressed. Not only wheat, but also barley and oats should be harvested into sheaves and kept in shocks. This method is of course redundant today, grain being taken into storage immediately in order to reduce spoilage and loss. Few gentlemen, Billingsley claims, will wish to attend markets and fairs for the necessary buying and selling, with its 'hacking and chaffering for a few shillings'. Instead, the bailiff must attend and be 'prime minister' there. Some of the work of shepherds and other 'servants' is detailed, then the sowing and harvesting of corn.

Beginning to conclude, he again stresses that 'nothing contributes more to the improvement of a county than good roads'. Where Billingsley was commissioner roads across Mendip were set out straight and broad (as much as 40 feet wide with verges), giving the erroneous impression today of their being roman roads. And finally, he reiterates the prime importance of good accounts. Towards the end of 'Waste Lands' he is again remembering the things he had intended to include: his organisation of material is less careful here than previously, presumably due to pressure to finalise the manuscript.

'Waste Lands' has given incredibly detailed advice on how to proceed with Enclosure in an exposed area. Not just the actual enclosing itself, but also the cultivation and all things necessary to establish a successful farm. Anyone travelling across Mendip today, where Billingsley was personally most active, will see the results of these methods laid out in front of them.[11] In terms of stock, the sheep that were predominant in his day passed (largely within his lifetime) from being kept mainly for wool to being intended for meat, then were replaced by cattle in the nineteenth century. For many years dairy cattle were in the majority, with almost no sheep kept in the area, then post world war two dairy has gradually become less profitable and any cattle now are more likely to be beef. Relatively recently the pendulum has swung back and sheep have once again become more popular. The changes of stock have mainly been due to market forces rather than the needs of the land. It is interesting that Billingsley has concentrated his advice almost entirely on the soil and crops, rather than stock. Management of the soil was his prime objective, from which all else would follow.

While today's farmers are generally complimentary as to Billingsley's methods, some of his ideas and practices are not seen as useful. One for which he is now criticised is the clearing and levelling of ground covered in mining spoil – this will have spread the pollution across the soil. We now know that lead is not only injurious to stock, it also pollutes the land permanently. Today, one farmer of land (once Billingsley's own) where there was previously lead mining, reports that his family had discovered many years ago that their cattle must not be allowed to graze the grass short as it affects their health. The grass itself is not the problem, it is the soil which they can pick up with it if they graze too low. Otherwise, the farm is very productive over 200 years since Enclosure; it is mainly permanent pasture though some fields can be used for arable. It is typical in that it was a dairy farm for many years but now focusses on beef and sheep.

Another method for which Billingsley is criticised today is the extent of his ploughing, dragging, harrowing and so on – up to four repeated ploughings. Today such aggressive intervention in the soil is frowned upon by many, ploughing being very much less popular than even five years ago, (and ten years ago his ploughing methods would not have had much negative response, apart from the expense). It is now believed that soil needs to be protected from interference wherever possible. A few farmers locally have converted to 'no plough', the equivalent of small scale 'no dig'. Some feel it is arguable, though, that Billingsley's thorough preparation of soil, including the repeated ploughing, was necessary for the original reclamation. Others feel it is the stony soil and bed-rock that makes ploughing impractical. Perhaps this is why Billingsley so favoured shallow ploughing. Some of his preoccupations with respect to the type of plough and use of oxen or horses are now obviously irrelevant with the advent of tractors. Billingsley himself was critical of ploughing when used to overcrop the land, but saw it as essential for reclamation. He advocated the liberal use of dung: today natural fertilisers applied frequently and with minimal ploughing are seen by progressive farmers as the way forward. The widespread overuse of chemicals from the 1960s, scaled back only recently, has had a huge effect on the soil and insect life, to its considerable detriment. We are only now learning the way to recover.

Most of the Mendip plateau, and all the land there Billingsley himself once owned, is now within the boundary of the Mendip Hills National Landscape (formerly an Area of Outstanding National Beauty). It is managed to retain its special character, which was created in part by the Enclosures. Many farmers there have recently taken up regenerative farming methods, encouraging eco-diversity. In some ways modern farming is returning to the methods used previously, such as those Billingsley advocates. Farming in a protected landscape means they have the opportunity to acquire dedicated funds for this, which many of them have taken advantage of. The current need for economic diversification has also led to a variety of non-agricultural activities on the land.

The system of cultivation, and the crops on Mendip, may today be different – and importantly there is a lot less arable there than in Billingsley's day – but much else remains of the landscape he formed. Permanent pasture predominates. The straight roads, together with the large and regular field shapes, shelter belts of trees, stone walls and stone-built farmsteads are all immediately noticeable. Looking a bit more closely, many of the stone lined pools for stock to drink from still exist; farmhouses do indeed face east or south, and usually have a barn on the western side (whereas today we would probably prefer the back of the house to be south facing). This landscape is his legacy. But the methods given in 'Waste Lands' have resonance very much more widely than in just this area. The agricultural wheel keeps turning, on Mendip as elsewhere: hedges go up, come down, then go up again; sheep predominate, lose out, then predominate again; dairy cattle appear, then disappear, and so on. The turns of the wheel are partly due to the economic situation and partly to what is understood – or fashionable – for agriculture.

Critiques of 'Waste Lands' contemporary with Billingsley are difficult to find. His colleagues in the Bath and West Society were simply appreciative, with no overt disagreement as to his Enclosure methods. By this time in his life he was seen by most of them as *the* expert. There was a clear national debate at the time with respect to Enclosure for agriculture, in addition to vigorous discussion on the financial and practical farming aspects. Attention slowly began to focus on its effect on the poor. Arthur Young, for example, having previously been an enthusiastic supporter of Enclosure, changed sides, becoming critical of the Enclosure movement.[12] So by this time he and Billingsley were not in accord, although neither seem to have been specifically critical of the other.

Today there is relatively little discussion on the effect of Enclosure on agriculture and very much more on its consequences for the poor. Perhaps for the same reason there has been very little recent academic attention to or critique of 'Waste Lands', apart from some comments from Williams, who recognized it as an important source of agricultural practice. Billingsley's agricultural reputation still rests primarily on the *General View…* with 'Waste Lands' largely forgotten. As for today's Mendip farmers, the overall response to his paper is that it is interesting to consider how much remains of his techniques and influence, that the methods he advocates are generally good, practical and – in particular – very thorough, easily acceptable and well suited to the vicinity. One local farmer remarked that he is grateful to Billingsley for having transformed the land and made it available for agriculture.

16
Earl Waldegrave's Steward

The Waldegrave family's Somerset estate included land on the Mendip plateau, close to Billingsley's own property there. For more than fifteen years he acted as steward and agent for them, from about 1788 to 1805. He continued in the role through his explosion of activity in other fields; and through various illnesses, house moves, writing, and periods of distinct stress and overwork. It immediately begs the question why he chose to take on the relatively onerous stewardship? It is possible it was for money, maybe when he felt financially stretched, although evidence shows that he did not press for his remuneration. He must have had some other motive. Perhaps it gave him further scope to hone his skills in a landscape he knew well, contiguous to his own. It did not end well.

His position as steward involved managing land, collecting rents and keeping accounts for an extensive estate covering several manors plus many other lands, properties and mining interests. Although it was paid employment it was a relatively prestigious role, certainly not that of a 'servant' in the usual sense, especially as Billingsley's employer was an earl of considerable fortune. George, the 4th Earl Waldegrave, had appointed Billingsley as his steward in 1788 but died the next year. George was succeeded by his son, also George, the 5th Earl, aged five at the time. The latter drowned in 1794. He in turn was succeeded by his younger brother, John James, the 6th Earl, then aged eight. In their minority the boys had guardians, in John's case his mother, Elizabeth Laura, Dowager Countess Waldegrave and three others.*

---

* Elizabeth Laura, Countess Waldegrave (1760-1816), led an interesting life especially when younger, but also saw tragedy. When she was six her mother (who had recently been widowed) secretly married King George III's brother, Prince William Duke of Gloucester – an act which provoked the King to fury and precipitated the passing of the Royal Marriages Act. However, the family remained at court. Elizabeth Laura married her cousin, the 4th Earl Waldegrave at the age of 24, and subsequently lived mainly in Somerset. She was still often at court, although during her relatively brief marriage she bore five children. She was widowed in 1789 when only 29; subsequently losing two sons to drowning. In 1797

Billingsley usually dealt directly with the countess, discussing the various projects, advising and presenting her with the annual accounts for her to inspect and sign. They seem to have got on well: during most of his years of service she apparently spoke very appreciatively of his work, as did the other guardians.

None of the Waldegraves seem to have shown any interest in agriculture despite their ownership of the estate; none of them were patrons of agricultural societies or even corresponding members, though this was a popular interest among aristocrats at the time. It is not known how Billingsley's appointment came about, but he and the 4th Earl will already have been generally acquainted, through Billingsley owning the neighbouring manor of Hazel, and through their mining interests and mutual contacts. Perhaps his reputation in agriculture preceded him. The Waldegraves will have relied on Billingsley for his management of the estate. In working for them he was occupying an enviable position, at a time when rank and privilege was so much more important than now. Stewards for such families were often highly regarded, as, for example, was Thomas Davis of Longleat, steward to the Earl of Bath. This could have been a factor for Billingsley in taking the position, although he himself was already highly regarded in his own circle so the stewardship would probably not have enhanced his reputation noticeably.

Over the years of his stewardship Billingsley not only managed the existing estates, already substantial when he joined them, he also helped to expand them, helping in the acquisition and supervision of land, mines and other property. For example, he was very active in helping to procure the Enclosure act for Chewton Mendip involving 1,500 acres in 1800 (in the face of much local opposition which had to be carefully handled); he subsequently reclaimed the land with his usual thoroughness. The improvement of land and the building of Chewton Mendip's Wigmore Farm for the family, widely regarded as a model project, was discussed in chapter 15. Billingsley also arranged the purchase of land at Radstock; was involved to some extent with the Waldegraves' mining interests there; and purchased the land and manor at East Harptree for them. This last purchase turned out to be the most problematic for him, as explained below. Despite seemingly good relations between Billingsley and the family over many years, at some point around 1804 or early 1805 things began to unravel.

Strawberry Hill House passed to her. In 1803 she bought Harptree House, together with 1,200 acres, intended for her son John but in her own name as he was still a minor. She lived there for many years thereafter, but also visited Strawberry Hill and died there.

John James, the 6th Earl, reached his majority on 1 Aug 1806.* The young lord immediately demanded a copy of his accounts up to date, which Billingsley 'hastily prepared'. There appeared to be a discrepancy of £1,303 18s 11d, owing to the earl (an exact sum repeatedly referred to in the dispute that followed). The latter required that it should be paid at once, in full. Billingsley admitted the discrepancy, but this, he countered, was because it had been only a partial account. He declined to pay straight away as, he said, the earl in fact owed him a good deal more money than that, mentioning about £2,000 or more. In the face of further demands Billingsley dug his heels in. Events then escalated into a series of court cases, starting in autumn 1807 and still ongoing at Billingsley's death.

The first legal case against him was brought in the Court of Chancery by the countess some time in 1807.[1] No record has been found of the outcome. It was rapidly followed by a second case: in the autumn of that year the young earl brought his own first case against Billingsley ('Waldegrave v Billingsley, 1807').[2] Both plaintiff and defendant made written statements for the court, from which we gain a little more insight as to what the problem was about. In the 'Bill' (the plaintiff's argument) Waldegrave claims that, in his capacity as steward and receiver, Billingsley owes him £1,303 18s 11d, that he had been asked for it several times but refused to pay. Although Waldegrave had offered to receive the money without prejudice the defendant had refused. Waldegrave complains that 'the account given is false and contains many overcharges, false charges and impositions', that Billingsley at times admitted that money came in, 'then denied it, sometimes pretending that he did not receive any monies'; he is accused of 'conspiring and confederating with divers persons at present unknown'; that he 'pretends' to have given Waldegrave 'extra services' but has not done so; that the defendant never did such services beyond what is commonly done and claims so to evade paying what is due. The earl goes on to say that Billingsley took possession of allotments and the proceeds therefrom, removing crops from the land to an adjacent farm of his own; kept money from the proceeds; that he has in his custody all the title deeds, rent books, books of account, day books and so on and refuses to give them up – and more to the

---

* The Countess' son John James, the 6th Earl (1785-1835), joined the army as soon as he left Eton, serving in a number of infantry regiments and rising to the rank of Lt Colonel, taking part in the Peninsula War and the Battle of Waterloo. In 1815 he married his long-term lover, Annie King (daughter of his regiment's chaplain), by whom he already had several children, and had four more thereafter. He inherited Strawberry Hill in 1797 through his grandmother, who was a Walpole: he and his son subsequently squandered the fortune between them. He died in 1835, after a relatively undistinguished career.

same effect. Grave charges. Even allowing for the tendency of legal language to make things sound far more serious than they were likely to actually be, this recital certainly reads as if Billingsley was actively defrauding the Waldegraves (the words 'pretending' and 'false', for example, occurring frequently).

Billingsley's 'Answer' to the bill (the defendant's argument) is also lengthy. He begins by accepting that there does appear to be such a discrepancy. He claims that as his figures were 'hastily prepared' it resulted in a partial account, describing in great detail the various transactions involved, including how much he was paid for his work as steward. Over the fifteen years this ranged from £50 to £100 per annum, he says, never more. This was not a particularly large sum for Billingsley, adding to the notion that his work as steward was for interest and/or prestige rather than recompense. But when other expenses are taken into account the discrepancy is actually, Billingsley claims, £1,096 1s 1d in his favour – and the real amount is far greater even than that as he has not demanded many of the expenses rightly owing to him (again, these expenses are set out in detail). He has had an accountant (named) check the figures from 1792-1805. This revealed some mistakes (given in detail) including sums relating to the coal works in Radstock and expenses for meetings there. Billingsley says he was 'in the habit of taking notes in hand if it was not convenient for the tenant to pay' on rent day, leading to some of the discrepancies. He adds that about this time he had been 'afflicted with severe bodily disease which [he] thought would be fatal', so balanced the books to protect his family, as he was 'under promise' to James Walker [of Radstock] to honour the latter's expenses in the earl's service. Corroborating evidence can be found as to Billingsley's ill health at that time.*

Billingsley continues his 'Answer' by saying that he also left a statement of his own expenses so that his executors would be able to charge them. However, he admits to having entered the £250 for Walker twice and says that amount should be added to the £1,303 18s 11d, making a total of £1,553 18s 11d. And he *did* perform 'great and extraordinary service' in addition to his normal duties. He gives three examples of this: enabling the Enclosure act for Chewton Mendip; his help in purchasing and overseeing the mining interests at Radstock – at 'the express request of the Countess' – and his purchase and management of the manor of East Harptree for the family. It is very probable that this last was the real trigger to the legal suits, although it is not mentioned by name in Waldegrave's statement, the land appearing only as anonymous

---

* For example, following publication in 1805 of his paper 'On the Utility of the Bath and West Society' (which was arguably not up to his usual standard), the editor of the journal commented that 'being taken ill… his health suffer[ed] considerably'.

'allotments'. Billingsley repeats that he has not charged for all his expenses, but was verbally assured of his reward and had waited for it until the young earl came of age.

Toward the end of his statement Billingsley quotes from the countess' letters to him, showing her attitude to the situation just before things became out of hand.* On 14 October 1804 the countess had written to Billingsley to say she had received a hint that he thought of giving up his service as steward:

> my confidence in your judgement and the zeal that you have always marked for his [the Earl's] Interest [...] what a serious loss I should consider it, you giving up the management on my son's concerns in Somersetshire. When I speak of myself individually that these are not my sentiments alone [?] I would equally lament with myself your adopting such a resolution [?] in all that concerns my estate in Somersetshire and especially about the purchase of East Harptree points out to me how much I am indebted to you. I acknowledge that Mr Coles has advised unnecessary delays from his questions as to the purchase [of the Manor of East Harptree] on account of the flaw that marked a vexatious kind of discomposure that may have hurt you. I think on reflection you will not increase my perplexities in leaving me involved as I am to work alone. Pray accept my most grateful thanks and pray be assured Sir, I am your dutiful and humble servant Waldegrave PS I hope Mrs Billingsley and daughter are well.

The countess' tone in the letters is noticeably friendly and positive towards him; she seems regretful, though the letter could also be construed as conciliatory. Things were still reasonable, with her at least, at this stage. Billingsley goes on to claim that he purchased the manor of East Harptree at auction, for £35,000, for the benefit of the Waldegraves but in his own name, and paid the deposit of £2,000 from his own pocket, signing an agreement which bound him to pay the rest by the December following. Another letter describes the manor of East Harptree to be 'subject to a contingency in the Title' (the 'flaw' mentioned by the countess) a matter which, if not favourably resolved, could have cost him £20,000. This probably explains the delay in purchase and difficulty in raising the money, as recounted by the countess. Problems in raising the full finance (due to the 'flaw' in title) and delays while the 'Guardians and friends' considered the purchase meant he then had to mortgage one of his own properties for £5,000 in order to fulfil his obligations.

---

* Unfortunately, the document is damaged – crucial parts being unreadable so that we do not have the whole text; however, the words quoted here are perfectly clear in the original, with a question mark added to the small sections in which words are illegible.

He had made the purchase himself, he said, because the earl was underage; the guardians had approved of it; both the countess and her son had been 'anxious for it'; the young earl had even attended the auction with him. Billingsley claims that the manor would now be worth £50,000 if sold, yet he has never received any reward or gratuity for this service – not for the purchase, survey, or management and improvement of the land for three years. Also, about twelve years previously he had 'engaged in the canal' which now benefits Waldegrave by several hundred pounds per annum, but from which he himself will sustain a loss equal to his whole salary for all his years of stewardship. He finishes his statement by repeating that he acknowledges the mistakes in the accounts; and that he holds no title deeds, or accounts apart from the day books (his working accounts) and that he wishes only to be renumerated with 'reasonable costs and charges'. A principled position.

Following this the earl applied to amend his statement to the court, which is recorded in the Close Roll Reports as allowed on payment of £20.[3] The subsequent judgement is known both from this report and from newspaper articles of the time.[4] The latter describe the situation briefly, continuing by saying that following the statement from the plaintiff a motion was made that Billingsley should pay the £1,303 18s 11d into court. The latter agreed to do so, although he resisted the charges and brought a counter-claim, addressing the chancellor in person, basing his argument largely on the facts as given in his statement above along with some additional information. He claimed that over sixteen years of superintending the estate he had expended £800 in travel expenses (for three manors at a distance from his home) and given care and attention beyond what could be reasonably expected of him. He had purchased the manor at East Harptree at the express desire of the countess, Lord Thurlow and the Marquis of Bath, who had repeatedly acknowledged their thanks for his services and intimated he should be rewarded. Through his efforts the estate had then become infinitely more productive than hitherto. And though the chancellor called on him for the balance yet he certainly expected that renumeration should be made him for his trouble and expense: he believes himself to be 'entitled to a fair and adequate reward'. The tone here indicates that Billingsley is indignant, as well as 'hurt' financially. The chancellor then observed that as Billingsley settled the accounts annually he should have made charges for this at the time, he did not approve of an account being held over for sixteen years and was apprehensive he could not relieve Billingsley under the present motion. He had no doubt the estate was improved but his decision must be governed by principles of public benefit and expediency.

The chancellor said he would read over the papers again before making judgement – which he was confident Lord Thurlow would not have bothered to do if the case had come judicially before him (Thurlow was a guardian of the young earl and had been lord chancellor until his death in 1806). One wonders whether the chancellor's attitude indicated that he felt the law was not doing Billingsley total justice. Although very critical of Billingsley holding over his claims so long, he is also helpful to him in re-reading the papers (perhaps to find some way Billingsley could succeed?). He then proposed (helpfully) that Billingsley bring another court action, which resulted in the Taunton bill, discussed below. Judgement was reserved, but the court ruled that within the next three weeks the original sum, plus the £250 Billingsley admits he had charged twice, should be paid by him to the court accountant, who would invest the money in trust in bank annuities with interest. Later information reveals that £2,285 4s 4d was invested in 3% consols in trust, on 2 Aug 1808.*[5] By this stage Billingsley must have felt his reputation was at stake, determining to vindicate himself.

The action now moved to Taunton Assizes, where in April 1808 three cases were heard by the court in one day. Events will no doubt have been followed closely by the local population as both the Waldegraves and Billingsley were well known in the district. The only records available are from newspapers.[6] The first case was brought by the countess, against 'Billingsley and Others', the mining partners mentioned earlier (chapter 11). This case was an action on a breach of covenant over a mining loan. Here there is a verdict, in the countess' favour: she was awarded £50.

In the second case Billingsley sued the countess with an action to recover compensation for the purchasing and subsequent management over three years of the manor of East Harptree. The newspaper states that while Billingsley was acting as steward and receiver (at an annual salary) he purchased the estate for Earl Waldegrave, but it appeared in evidence that the countess was enjoying the benefits of it. Therefore, the jury returned a verdict in favour of Billingsley, for £300 plus costs.

The third, most important, case was brought by Billingsley against the earl. The newspaper reports that this action had been helpfully proposed by the chancellor in his Court of Equity the previous year. The current case was to recover remuneration for Billingsley's extra services over eight years in cultivating and improving a 600-acre estate of waste land on Mendip for which his lordship enjoyed £500 per annum, but which before Billingsley's services

---

* The total required to be paid in by Billingsley was £1553 18s 11d; there is no explanation as to where the additional funds came from. Perhaps from Waldegrave.

had been entirely unproductive. Plus, Billingsley performed other extra services during the earl's minority which were outside the normal scope of a receiver or steward, for which he was verbally assured he would be liberally rewarded once the earl came of age. Waldegrave, however, made 'Defence of Infancy' (that he was a minor at the time) and was therefore not liable in law. This was unfortunate: should not Billingsley's lawyers have expected the 'Defence of Infancy' and brought the action against the guardians rather than the 'Infant'? But Billingsley also provided evidence of some other 'trifling services' since the earl came of age and the court found he was certainly entitled to recompense for them. This verdict was also in Billingsley's favour: £40 plus costs.

The next case returned to the Court of Chancery, brought by the earl in 1808 ('Waldegrave v Billingsley and Others 1808').[7] This was a return to the mining case begun by the countess. The 'Others' named were George Pope, John Saunders [Tudor] and John Taylor, (two of these being gentlemen of East Harptree; the third can be identified as the 'JS Tudor' who kept financial records for the Waldegrave's estate, although for some reason his last name is not entered in the court documents.) These men were presumably thought by the earl to be among the 'divers others' with whom Billingsley was 'confederating'. The Waldegraves had already obtained a court order stopping the partners from mining as of 1 July 1805.[8] The original lease stated that they should abide by the 'Ancient Laws of Mining' and in case of dispute as to the 'quantity of recompense' would be judged by 'two Indifferent persons, one to be named by the occupier...and the other by the said Defendants'. However, it seems a reason other than recompense was chosen as the subject of the suit, so they were not judged by the ancient laws of mining but in a standard court of law.

Interestingly, Tudor's accounts for Waldegrave are available only until December 1805, after which they cease.[9] Whether this means that Tudor resigned, or that his employment was terminated, is not known. A page from his accounts, dated 6 November 1805, shows that 'John Billingsley Esq, [was] Steward then', suggesting that Billingsley did not remain steward for long after that date. The complaint behind this suit has been described earlier (in chapter 11), the case resulted in the injunction against the partners which was to remain until the court made a further order. Unfortunately, no record of the final outcome of the 1808 suit has been found.

By this point there is definitely a sense that the earl was pursuing Billingsley, using the law in any way possible. The 1808 case was soon followed by another ('Waldegrave v Billingsley 1809', with a 'Revised Bill').[10] Here Waldegrave rehearsed very similar grievances about more general estate matters, causing Billingsley to go back as far as 1802 in his reply. Rather than the original sum claimed by Waldegrave, Billingsley again counterclaimed with what he believed was due to him from Waldegrave, plus £20 per annum for twelve years and a further £60 16s 1½d, laying out the reasons in some detail. On 27 November 1810 the two sides agreed to refer the matter to arbitration. Each side appointed a referee, but the process took some time.* Before the arbitrators had finished deliberating Billingsley died, on 26 Sept 1811. On the 28th of that month, only two days later, the arbitrators arrived at a decision, and that due to the death proceedings should cease. Unfortunately, the papers detailing the arbitrators' verdict do not appear to have survived, so we have no information as to their reasoning, but their main conclusion must have been in Billingsley's favour. Court proceedings later showed the arbitrators recommended the court should order that £582 16s 11d interest from the trust be paid to Billingsley's executors, the rest of the cash in the account (less than £110) to go to Waldegrave, with the parties paying their own costs.

Within weeks Waldegrave applied to revive the case: 'Waldegrave v Billingsley 1811', now suing Billingsley's executors.[11] The executors for their part, claimed the case became 'abated' (interrupted) on the death of Billingsley, that they should have the stock purchased and the £582 16s 11d interest and wanted the case against them dismissed with reasonable costs and charges which have been 'most wrongfully sustained'.

Waldegrave's side reiterates the now familiar arguments that Billingsley owed for sundry items and that the money deposited with the court should go to them. However, the court decreed that the stock should be sold: it now amounted to £2,464 7s 8d (although it seems Billingsley had paid only £1,553 18s 11d into the court). The court decided, using the advice from the arbitrators, that the money arising, and the interest plus a dividend making £657 16s 11d overall, should now be paid to Billingsley's executors. The £99 16s 3d in cash from the trust was due to Waldegrave, according to the act of parliament relating to such orders.[12] Waldegrave had lost.

Thus Billingsley spent the last years of his life under the cloud – and with the attendant anxiety – of legal processes. Even if the sums involved, although not inconsiderable, were not crucial to him, any judgement against his integrity certainly would have been. It seems inconceivable that he would lie to the court, yet he repeated all his claims in person more than once. It is entirely possible, as was suggested at the time, that the on-going litigation prompted sufficient stress for another, fatal, asthma attack at a time when his health was said to have been much improved, although this is conjecture (see the final chapter for further

---

* Mr James Camper Wright was nominated for Waldegrave, and Mr Thomas Grame for Billingsley.

discussion of this). The saga reveals several aspects of Billingsley's character. Arguably it was unwise for Billingsley to go head-to-head with a family such as the Waldegraves – resisting such powerful people, who could hire the very best lawyers to work against him, at a time when rank and fortune meant so much. His reputation could easily have been ruined even though he was innocent. He could have submitted at the first hurdle, yet he stood up to them.

The series of cases must have cost both Billingsley and Waldegrave considerable sums of money (one is reminded of Dickens' 'Jarndyce v Jarndyce'). Billingsley had expected to be rewarded but did not demand it at the time, presumably believing that an honourable lady or gentleman would pay his expenses and reward him for his efforts in due course, being reticent about demanding payment. He could, then, be accused of being overly trusting. But the case also shows – and not for the first time – that on occasion he could be muddled and inaccurate, most noticeably at times of stress, overwork and ill-health. And despite his life-long emphasis on careful accounting he could be casual about payments, either to others or himself. It is worth noting that he used his illness as an explanation, not as an excuse, yet it might be considered excusable at least to some degree. Perhaps he had just taken on too much – he may have had problems saying no – which again would explain his being overloaded.

For a non-lawyer the legal aspects are not transparent. Reading through the depositions and judgements there does seem to be a tension between legal and natural justice. Also, considering his length of service, the family's trust in his handling of significant financial affairs over that period of time, and his obvious willingness to help in difficult circumstances one does question whether some other solution might have been found. Perhaps the young earl sued for reasons other than simply to recover what he claimed was owed to him. His underlying motive for the repeated cases is not obvious – there is a great temptation to speculate (resisted here with difficulty!). Billingsley was hardly cold in his grave and his family still in the first stage of mourning when the earl repeated his relentless pursuit of the money owed. There is a sense that a gentleman of integrity and of substance would at least pause at Billingsley's death, if not dismiss the supposed debt at this time. It was an unsavoury business. It seems, though, that the case had absolutely no effect on Billingsley's excellent reputation. His friends were noticeably sympathetic, and even though it must have been hot news at his death it has been largely forgotten. It never features in subsequent discussion of his capabilities, achievements and integrity, which remain unsullied.

17
LAST YEARS AND LEGACY

DURING THE FIRST few years of the new century Billingsley was still working extremely hard. His work in Enclosures and drainage was very much on-going, with some activity in mining and committee work for both canal building and, of course, the Bath and West Society. It is clear, though, that from the beginning of the 1800s the pace of his activity was beginning to slow: the second edition of the *General View.*. had been published and the renovation of Ashwick Grove was completed, the mansion soon to be rented out.

Two views of Camden Crescent, Bath (2018)
Billingsley's home was No 18, with the white door surround

Going back to 1796, as mentioned earlier the family had been temporarily resident in Bath, at no 16, Argyle Buildings. It was in a central location, but it was a small house with modest accommodation for a family who will have had a number of servants and been used to having eight bedrooms and an estate. They do not seem to have stayed there long. It is probable that Billingsley's uncertain health at that time was the reason for this first stay in Bath, so that he could 'take the waters' as was fashionable then. It was usual in those days to rent in Bath for the 'season', as maybe the Billingsleys did from 1796 on, as they seem to have divided their time between Ashwick Grove and Bath.

It was probably in the autumn of 1803, when Lydia White took over the tenancy of Ashwick Grove, that the family moved back to Bath full time. There is no definite information as to what prompted the move, only a suggestion that the climate at Ashwick Grove was not conducive to Billingsley's health: relocating to Bath will have meant he was able to get out of Ashwick's damp climate. The move was probably not primarily for society reasons: Billingsley already had a very wide circle of contacts and his only daughter was still a child, so not yet in the marriage market. Bath would also have been more convenient for contacts and business meetings than rural Ashwick. In 1803 he became a member of the Bath and West's newly established 'Superintending Committee', which stipulated that members should live within five miles of the city.[1]

Although they must have lived somewhere temporarily when they initially returned to Bath, but their main home was at no 18, Upper Camden Place (subsequently renamed Camden Crescent). This house is part of a typical Georgian crescent, a handsome row of houses built of Bath stone.[2] The architect and builder was the well-known and highly thought of John Eveleigh. The row is dated 1788 – the first of the major projects Eveleigh undertook in Bath, now grade I listed. There were originally 33 houses, no's 16 and 17, the central pair, were topped by a pediment with the arms of the 1st Earl of Camden, Charles Pratt. No 18 occupies a prime position immediately to the right of the central pair. Being built on an escarpment means the row has magnificent views across the city of Bath, something that must have been attractive to Billingsley, especially as his own home at Ashwick Grove had no view. In 1789 (before he had arrived at Camden Place), while the houses were not yet quite completed, the ground gave way and the east end of the row suffered a landfall. Several houses collapsed, never to be rebuilt. This meant that the pediment was no longer in the centre of the crescent, but otherwise did not affect no 18, the remaining houses being successfully stabilized.

The family was in residence by the end of 1805.*[3] They remained there until at least early 1808. Soon after that Billingsley was again back at Ashwick Grove, apparently believing that his health was improved once more. In 1809 Bath suddenly became very much less fashionable, as the Prince Regent relocated to Brighton for his leisure activities, so fashion too may have played a part in the decision to move back to Ashwick Grove.

Billingsley took on few new tasks after he returned to Ashwick. His only daughter Marianne was then fourteen years old – perhaps disappointed at losing the excitements of Bath. During his time in Bath Billingsley had had the pleasure of being awarded the Bedfordian gold medal, among other honours. An engraving, thought to have been made about that time, shows a rather more youthful looking man than his only known portrait in oils, but with the same intelligent and benevolent gaze. On the downside, during his time in

---

* The secretary of the Bath and West, Nehemiah Bartley, wrote that he 'called on Mr Billingsley at Camden Place on 27th November, 1805'.

Bath, Billingsley had had the difficult job of negotiating over the Nehemiah Bartley affair. But very much worse would have been the ever-present problem of the Waldegrave court cases, which continued (as described in chapter 16) from 1807 until his death.

While in Bath he began to re-arrange some of his business affairs. He re-mortgaged his land at Hazel manor and Ubley, this time with Thomas Jolliffe of Ammerdown, who provided £8,000.[4] His last will was made on 16 November 1807 (while still at Camden Place). Then in September 1810 he withdrew from his co-partnership in the Oakhill Brewery.[5] These are signs that he was putting his affairs in order. Even so, he continued with a punishing schedule. Then on 26 September 1811 he died: at 'his seat at Ashwick-Grove'. His death seems to have taken everyone by surprise. According to a note in the *Bath Chronicle*, Billingsley

> had for many years past laboured under an asthmatic complaint, which had several times so far increased upon him, as to alarm his nearest friends with fears of his speedy dissolution. But he had lately recovered from one of those attacks, with signs of renewed health, from which his far longer continuance was hoped.[6]

An asthma attack can occur very suddenly without warning, and in those days there were no reliable medicines to ameliorate its effects. The anonymous author of the above article goes on to say (alluding to the Waldegrave court case) that

> it must be peculiarly gratifying to his family and friends, that an unpleasant dispute between him and the Waldgrave [sic] Family (probably occasioned by the wrong designs of others) was terminated, by an able arbitration, in a manner most honourable to his conduct and memory, two days after his decease.

He was buried at Ashwick St James on 3 October 1811. A large marble tablet was soon erected inside the church by his grieving widow, with fulsome praise for his talents and character.[7]

In his will, Billingsley left everything in trust (apart from some personal and household goods). His executors and trustees were named as Richard Perkins (long-term friend), Thomas Parsons and William Parsons (nephews, though William had predeceased him) and Edmund Broderip (friend and solicitor). Probate was granted 16 November 1811.[8] All was to be sold as soon as convenient, the resulting funds to be put into good securities. The interest was to go to his wife (provided she remained unmarried – a usual provision at that time) until his daughter Marianne attained the age of 21. Thereafter his wife was to get only a moiety of the interest for life, the rest to go to Marianne, and after his wife's death Marianne was to get the residue and remainder; if Marianne died without issue then his nieces and nephews were to get the residue and remainder. His wife was also to get the 'provisions and use of plate etc' during her lifetime plus £50 within two months. He made a number of small bequests, including to his nephews; £20 a year to Mrs Jordan for life (the widow of his partner at the brewery); and £20 to Miss Twigge, Marianne's governess. A conventional will, but it took many pages to set out. The only unusual clauses indicate that the on-going legal case with the Waldegraves was on his mind when it was drawn up: 'there is an account remaining open between me and Lord and Lady Waldegrave and a Suit now depending in the Court of Chancery... I hereby authorize my Trustees to act on that as they think fit and proper, whether by Arbitration or any other means'. Unfortunately, though, from his will we get little idea of the extent of his fortune: it makes (as was usual) only general mention of his 'lands and hereditaments' and other assets and how they should be disposed of, with no detail as to what they comprised.

When the Bath and West Society next met after his death the mood was sombre. As the minutes of the October general meeting showed no sense of surprise his death must have already been general knowledge within the society by the time of that gathering. Sir John Cox Hippisley made the main speech in his honour, to their 'highly valued, deeply lamented and never to be forgotten associate...several other Gentlemen followed suit' with their own tributes. It was unanimously decided that the society should erect a 'Monument of Veneration' for their departed friend to be displayed in the society's room. This, they agreed, should either take the form of a marble bust, or, if impractical, a painting by a distinguished artist. The only portrait of him known to them was owned by Mrs Billingsley, to whom application was to be made to copy it. In addition, a tablet should record his virtues, and an engraving made of the portrait and affixed to the next volume of the society's papers. Expenses should be defrayed by a subscription.⁹

Fellow member Mr Grame accordingly wrote to Mrs Billingsley, who received the request as an honour.* She remarked that 'if the artist is fortunate in taking a good likeness, I purpose having one in oyl [sic] for myself, as those in crayon are seldom lasting'. The engraving was made, but sadly has since been lost, the only record of it being the copy printed in the society's next

---

* Grame was Billingsley's chosen arbitrator for the Waldegrave case.

publication. The tablet finally erected bore the following inscription:

> Non sibi sed toti genitum se credere mundo *
> JOHN BILLINGSLEY, Esq.
> One of the original Founders, one of the greatest Ornaments, and for 32
> Years a most active and able VICE-PRESIDENT of this Society, whose Ardour
> in acquiring Knowledge was only equalled by his Delight in imparting it; and
> whose Zeal in promoting Objects of public Utility was as conspicuous as his
> Judgment in discerning, and his Ability in carrying them into Effect
> The BATH and WEST of ENGLAND SOCIETY
> In grateful Remembrance of his transcendent Merits have caused this Tablet
> to be inscribed.

A gift of £100 to the society was made by Mrs Billingsley, who said that they would have received it in due course, had her husband been able to make a new will as he had intended. There was further lamentation at the annual meeting on 17 December 1811, when the president, Sir Benjamin Hobhouse delivered the eulogium.[10] As it happened, Cox Hippisley had also died by the end of the year, but, as Kenneth Hudson pointed out, while the former's services 'were acknowledged in a modest fashion… Billingsley received a hero's treatment'.[11]

In his lengthy (and often quoted) eulogy Hobhouse began by observing that in the years before the establishment of the society:

> the principles of Agriculture were but little understood; that the implements of husbandry were in a very rude condition, and that the livestock of the Farmer was far distant from that perfection which it has since attained. This was the degraded state of Agriculture, at the period when Mr Billingsley, aided by a few other congenial and public spirited individuals, applied his powerful understanding to its improvement.

Hobhouse went on to remark that among Billingsley's 'rare and extraordinary talents…nature [had] endowed him with a powerful and vigorous understanding'. Excerpts from the eulogy also address other aspects of his character and aptitudes, such as his skill at negotiation: 'to heal differences, and restore harmony was the favourite pursuit of his mind'. He mentioned Billingsley's inevitable discomfort at the situation with the Waldegraves:

---

* 'To think that he was not born for himself alone, but for the whole world', Lucan.

'how his heart must have been rent by the disputes and litigation in which, towards the close of his life, he was unhappily involved'. He goes on to say that 'the immediate cause of his [Billingsley's] dissolution was perhaps that sensibility of mind which so strongly marked his character'. This comment might have passed unnoticed by those unaware of the circumstances, had he not followed it by further allusions to the situation. Hobhouse strenuously avoids being culpable of slander – he 'will not pronounce any opinion' – but there is a definite hint as to the ultimate cause of Billingsley's 'dissolution' (death). Then:

> had it pleased the Great Disposer of all events to have spared his life for a few days only, he would have seen that innocence, of which he was proudly conscious, clearly established: and that integrity, which he valued more than life, firmly upheld by the unanimous award of the arbitrators, to whom the final adjustment of the matters in litigation was committed.

The words from the eulogium subsequently most often quoted, though, are Hobhouse's declaration that 'He Enclosed Mendip! He Drained the Levels! He wrote the agricultural Survey of Somerset!'. This last statement is clearly accurate, but the first two assertions could easily be taken as meaning that Billingsley was the only or main person to do these things, whereas in reality in both cases he played a very important part as one of many.

More objective than either Mrs Billingsley's memorial tablet or Hobhouse's eulogy, was an obituary published anonymously in the *Gentleman's Magazine* shortly after his death.

> At Ashwick Grove, near Bath, aged 62 [sic], J. Billingsley esq. author of "The Agricultural Survey of the County of Somerset..." His life was one continuous round of active utility, particularly in agricultural pursuits: and for the general benefit of the country, in the promotion of inclosures. And the cultivation of vast tracts of waste lands; encouraging the cutting of canals and the improvement of public roads. In all such undertakings, Mr Billingsley was ever found at his post – able, intelligent and communicative; rendering his aid as a committee-man, or faithfully discharging the powers with which he was vested, and the confidence reposed in him.[13]

The *Bath Journal* of 30 Sept 1811, meanwhile, described him in another often-quoted phrase as a 'a Boast, benefit and Ornament to his neighbourhood'.[14]

*John Billingsley, late Vice-President of the Bath and West Society
Engraving, a copy reproduced in the Society's Letter Book*[12]

At the time of her father's death Marianne was only 17. On 12 February, 1814 she was married to the Rev George Penrose Seymour. Being then 20 she was still under-age so would have needed the trustees' blessing, but Seymour was a good match from a family with resources. A satisfactory pre-nuptial agreement was made.*¹⁵ After their marriage Mary Billingsley lived with her daughter and family. They moved about to some extent, living in the Isle of Wight, in Wraxall and eventually in Marksbury, Somerset, where Mary died in 1828. George and Marianne had twelve children. Only one of their descendants seems to have found any degree of fame: their son Arthur Penrose Seymour, who emigrated to New Zealand and had a successful career there in public life.†

But it was another of their sons, Henry Fortescue Seymour, who must have inherited what was left of Billingsley's agricultural prizes. One of the puzzles of Billingsley's legacy has been what became of the cups and medals he won from the Bath and West. They had apparently survived the general clearing out that disposed of his papers, as they had been lent to the society on several occasions for exhibitions in the early twentieth century, then disappeared from view. In the 1960s Robin Atthill traced both the silverware (some shown in chapter 7) and the only known portrait of Billingsley, to being then in the hands of a Mrs W Moger of Bath, Billingsley's great-great-grand-daughter.‡ This is the portrait shown at the beginning of this book. Before she died in 1963 she gave the painting to the Victoria Art Gallery in Bath.§ In 1814 the Bath and West reported that 'Mr King, statuary, had presented an admirable model of the late Vice-President John Billingsley Esq, for which gift he was awarded the Bedfordean Medal as a testimony of high respect'. Unfortunately, there is no record as to the whereabouts of this model.

Although Billingsley's will directed that Ashwick Grove should be sold as soon as convenient the family home was not disturbed immediately. There was time for the very efficient decluttering that must have taken place after his death: none of his papers have survived. No time was wasted in disposing of other assets. On 17 October 1811, within three weeks of Billingsley's death, his Ashwick farms' stock was advertised for sale by an auction to be held at Ashwick Grove, 'without reserve'.¹⁶ The items are listed precisely: '43 prime Leicester and South-Down ewes; 21 ditto chilver [female] lambs; 2 Merino rams; 1 ditto 2 ewes; 1 ditto chilver lamb; 3 dairy cows; 4 heifers; 1 bull; 2 draft mares' plus a variety of crops in-field; 90 tons of hay; about 40 acres of grass;

---

* The court case with the Waldegraves had already been finalised by 1813.

† Arthur Penrose Seymour (1822 -1923); in New Zealand from 1851; farmer; JP; member of house of representatives; first chairman of board of education [marlboroughonline.co.nz].

‡ Mrs Moger, born Mary Kathleen Fortescue Weigall, was a descendent of Henry Fortescue Seymour.

§ It remains under their auspices, but hangs in the Kaposvar Room at the Bath Guildhall, where it is on view by request.

waggons, harnesses; 100 fleeces and sundry other articles. No oxen are listed. The farm and flocks seem surprisingly small for a man with interests and funds like Billingsley. Possibly he had already sold off a larger amount of stock before his death, plus he may well have held further stock on some of his other land. Following the auction the farms were to be let, from Lady Day next.

Hazel Manor was also up for sale almost immediately.[17] The estate extended to '1007 acres, statute measure, of Arable, Meadow and Pasture and about 70 acres of Woodland'. It seems the estate was not attractive to buyers, whether due to price or location. A sale was not finalised until 1818 (as described in chapter 13). Selling the Ashwick Grove estate proved to be no easy matter either. Proceedings were not set in motion until April 1813, when it was advertised in the *Bath Chronicle*, evidently without success. It was advertised again in July, twice, and by late July it went to auction.[18] The sale involved the mansion and a total of 168 acres for what were known as Billingsley's 'Ashwick Farms'. These were made up of: 50 acres of land 'in hand' (including the mansion and associated 20 acres; the Fosse Estate; 2 cottages; park land; and plantations); two further farms totalling 40 acres ('Sims and Norris's with a limekiln'); and another 13 acres ('Longhouse Ground') all in Ashwick parish; plus land in Shepton Mallet, 12 acres ('Jenkyn's'); and in Stoke Lane, 41 acres (a 'good farm house', water meadows and pasture). These were all occupied by Robert Hues. No record of the auction can be found, but while the land was sold off in separate sections, the mansion and associated 40 acres or so failed to sell at the time.

It took until 1817 for the sale of the estate to be completed, to the Strachey family at a cost of £5,500 (as described in chapter 13). Some of his land must have been disposed of in other sales – for example, the majority of the manor of Stoke Lane. The problems in selling the various parts of Billingsley's estate may well have been affected by the difficult economic situation at the time, but his land and mansion were clearly not particularly desirable – despite the seemingly excellent condition of the land. Eventually all was sold, his daughter married, and the trustees' business wrapped up by 1818, apart from a few shares which were not sold until the next year.

Considering his origins, Billingsley had left a considerable legacy to his family. Judging from the torrent of tributes on his death his contemporaries certainly valued him for his attainments and capabilities. Many at the time (particularly those in the south-west) saw him as 'equalled by few and excelled by none'. But what of his reputation with later generations, into the present day? It is said that the enduring legacy of any individual consists of what they have written and the permanent change they have brought about. Billingsley was still being loudly praised a generation after his death for both his publications and his practical achievements. For example, in 1839 the historian William Phelps described the results of Enclosure on Mendip:

> The drainage of the swamps and cultivation of its surface, where it was capable of improvement, together with plantations made within a few years, have altered the aspect of the forest. Farmhouses have been built, and an increasing population is spreading itself; roads communicating with the different towns and villages round its base intersect this once dreary mountain in various parts, affording an easy access to it... Arable husbandry predominates wherever the soil will admit the plough.[19]

In only a few lines Phelps sums up Billingsley's effect on the land. Sir Thomas Dyke Acland also paid great attention to him, writing in a prize-winning essay of 1850 that his 'report to the Board of Agriculture...shows him to be have been far in advance of his time, in both his ideas and practice'.[20]

In general, though, after the mid-nineteenth century attention to Billingsley began to fade, probably at least in part because agriculturalists were preoccupied with their own improvements and the Board of Agriculture was replaced by another organization.* Since the mid-1800s mention of him has been infrequent. Among the few comments, in the 1960s Atthill emphasized the *General View's..* significance in describing Somerset during a period of transition, under the double impact of the industrial and agricultural revolutions. He saw this work as 'Billingsley's Great Memorial'.[21] Yet on another occasion Atthill himself described Billingsley's success in ploughing as his 'greatest single contribution'.[22] This illustrates a central difficulty in appraising his work: several different aspects could be chosen as his greatest legacy, depending on one's interest and point of view. Most of them, though, were linked by his passion for agriculture. At heart he was a practical farmer, the main tenets of his agricultural practice established surprisingly early and changing during his lifetime only in that they evolved into a coherent system. He was a leading 'Improver' of some significance, both in the sense of the word as then used in agriculture, and in the more general sense of making existing methods, implements and so on, more efficient and effective. But he was not an inventor. Even the double furrow plough, the item most often quoted as his creation, was not - as he himself pointed out - his own invention. And his efforts to introduce it were only very partially successful.

---

* The Royal Agricultural Society was established in 1838.

Today the most important of his many publications is still usually considered to be the *General View..*, its significance increased both through its being written, as Atthill recognized, during a period of change, but also by its being part of a national series. The best of the county reports – and Billingsley's was certainly one of the best – have remained the most useful source of information about agriculture nationally at that time. It is still seen as the foremost contemporary account of Somerset farming as it was at the end of the eighteenth century, with a shrewd look at what could be done to advance it enshrined in the 'Hints for Improvement'. Historians of agriculture still return to the *General View...* It was – and is – useful as more than just a description of how things were: many of the ideas and methods Billingsley discussed proved to be very influential during the succeeding years, often nationally but even more so locally.

As argued above, though, 'Waste Lands' stands alongside the *General View..* for its clear instructions on how to actually go about agricultural improvement, especially in the matter of reclamation of the land. His advice stems from personal knowledge and experience, which could be followed by others engaged in agriculture. It is *the* detailed practical handbook, all meticulously costed, of the methods which he himself used to enable large areas of previously infertile ground to be rendered fertile, and to remain so. Its messages are still appreciated by today's farmers and agriculturalists. As an improver, Billingsley stressed the importance of soil and ways to prepare it as being crucial, including methods for ploughing and when – or not – to plough. Similar discussions are taking place today, involving no-plough and regenerative agriculture, for example. Regrettably, some of the advice in his publications, though proven to be wise, has not always been taken, as in overcropping. And some of his most cherished ideas fell by the wayside even before his death, the double-furrow plough and use of oxen rather than horses among them.

Billingsley's most acclaimed twin achievements, the Enclosure of Mendip and drainage and Enclosure of the Levels, allowed huge agricultural change, the new improved landscape itself being his legacy. His work in drainage was surprisingly innovative. Without prior training or experience he seems to have had a natural understanding of what was needed, being confident in his own judgment, describing how, though his plans were 'ridiculed', he persisted with his ideas. He is sometimes praised for his work on the Levels and denigrated for that on Mendip – the latter for social rather than agricultural reasons. The Enclosure and drainage was a huge amount of work, which led to a significant increase in the farming capacity of Somerset. He claimed that the methods he suggested would have a huge influence on the capacity of the nation's land: it could add one-third to the produce and value. He often used the increase in profit as a temptation to others to Enclose, but for him this was not the most important factor. For him a greater crop of grain after Enclosure was more important than the associated increase in rent.[23]

Billingsley was also, of course, influential in a wide variety of other ways, including through membership of the Bath and West Society and improvement to the infrastructure of transport, through roads and canals. As another example, he is remembered in the village of Oakhill mainly for his part in the Oakhill Brewery, which expanded the village and gave it prosperity; it is at least as much for this as for his agricultural fame that his name lives on in the immediate area – despite it being part of Mendip. None of the above would have been achieved without his particular personality and exceptional personal skills: in communication, negotiation and vision. His financial success was due to his skill in speculation and attention to costing. Over the years his initial urge to accumulate money and land lessened once he was established, although it never ceased. Even early in his career he diversified from simply accumulating in the wool trade to becoming involved in turnpikes and the brewery, with profit seemingly not the over-riding aspect of either. Later in life he speculated mainly to pursue his interests, rather than for pure accumulation. He did not become a spectacularly wealthy man, or a major landowner: he changed his focus from acquisition to practice.

Like most men Billingsley had his failures. Some of his speculation failed, his investment in canals for example was only moderately successful. But from today's viewpoint probably his greatest failure was his inability to recognize the consequence of Enclosure for the poor and the commoner. This (and his attitude to the poor in general) has tarnished his reputation. There is a sense, though, that his conscience must have been pricked, as later in life he gave so much attention to the subject, trying to justify his position but ineffectually – as if it was too hard for him to change his mind, as Young had done. His attention was fixed on the agriculture, and in purely agricultural terms his Enclosure activity was very successful. Considering the situation in his day, when war and inflation meant the nation was in desperate need of increased food production at reasonable cost, it was a great achievement that he was able to set out reliable ways to provide it.

Historians today pay much less attention to Billingsley's actual agricultural practice than to its social effects. Discussion of Enclosure recently has focussed strongly on its effects on the poor: while agriculture and the economy gained from Enclosure, the poor and landless lost. But it was obviously not a simple equation. It is chastening to consider the results for the nation – the poor in particular – if the Enclosures then completed had not produced that extra food. Further research on what would have been lost by the one, versus the admittedly negative effects on the other, would be useful. Another aspect which would repay attention is why the Enclosures on the Levels were so much less contentious than those on Mendip.

Due to the gaping holes in the primary sources, telling Billingsley's story has involved collecting and presenting a huge number of small details, many of them relatively insignificant, yet often the only information available on a given topic. Piecing the bits together has been like assembling a mosaic. The picture (with some remaining gaps) reveals a complex man with many attributes and achievements. He was not a giant of his age, but he certainly had some stature as one of the many who together brought about the agricultural revolution. He undertook an astonishing amount of work in a wide variety of fields – with outstanding results. It is quite probable that, as Hobhouse hinted, the stress of the court case together with overwork ultimately caused his death. He may well have achieved even more if his life had not been unexpectedly cut short.

Billingsley described himself as a 'Speculative man turned Farmer'. Accurate, but modest. Having looked at his legacy a better description would be the well-deserved and more prestigious 'Entrepreneur turned Agriculturalist'.

Abbreviations

TNA	The National Archives
SWHT	South West Heritage Trust
SHC	documents held at the Somerset Heritage Centre of the SWHT
DHC	documents held at the Devon Heritage Centre of the SWHT
BRO	Bath Record Office, Bath and North East Somerset Council
MERL	Museum of English Rural Life, University of Reading
RSA	Royal Society of Arts

References

Preface and Introduction
1. Atthill, Robin, 1971, *Old Mendip*, David and Charles, pp45-54
2. Williams, Michael, 1970, *The Draining of the Somerset Levels*, Cambridge University Press
3. Williams, Michael, 1976, 'Mendip Farming: the last three centuries'; in *Mendip: a New Study* [ed Atthill, Robin), David & Charles, pp102-125
4. Hudson, Kenneth, 1976, *The Bath and West, a bi-centenary history*, Moonraker Press
5. Billingsley, John, 1794, *General View of the Agriculture in the County of Somerset with Observations on the Means of its Improvement*
6. Hutchinson, Joseph, *Portrait of John Billingsley*, reproduced by courtesy of the Victoria Art Gallery, Bath and North East Somerset Council

Chapter 1 Ashwick – Background and Context
1. SHC DD/SH/C1165/8, 1782, *Day and Masters Map of the County of Somerset*
2. Billingsley, John, 1794, *General View..* p7
3. Purdy, Frederick, 1860, *Journal of the Statistical Society*, vol 3, no 3, pp286-329
4. DHC 1262M/E15/8, 1774, *Correspondence re Estate (Miles to Fortescue)*, Fortescue Estate
5. Locke, Richard, 1790, *Letters and Papers of the Bath Society*, series 1, vol 5, p180
6. DHC 1262M/o/TSO/28, 1671, *Quitclaim, Manor of Ashwick, Henry James to Arthur Fortescue*, Fortescue Estate
7. DHC1262M/o/E/15/14, 1791, *Book of Reference to Map of 1791*, Manor of Ashwick
8. HC 1262M/o/E/15/3, 1760, *Map of Ashwick Down*, Fortescue Estate

9. Davis, Fred, 1996, *A Shepton Camera*, vol IV, Shepton Mallet Amenity Trust
10. Achinstein, Sharon, *Nicholas Billingsley* in Oxford Dictionary of National Biography
11. Murch, Sir Jerom, 1835, *A History of Presbyterian and General Methodist Churches in the West of England*, R. Hunter
12. Billingsley, Nicholas, 1721, *Rational and Christian Principles: the Best Rules of Conduct, with an Appendix containing a Vindication of the Author*
13. Woodland, Patrick, revised Benedict, Jim, *Hubert Stogdon* in Oxford Dictionary of National Biography
14. Stephen, Leslie, *Dr James Foster* in Oxford Dictionary of National Biography
15. DHC 1262M/E15/10, eighteenth century, *Surveys of Ashwick, Oakhill etc*, Fortescue Estate
16. ibid, n 15

Chapter 2 John Billingsley's Life – Synopsis with a Table of Key Events

1. SHC DD/FS/1/1/30-31, 1808, *Lease and Release of land at Masbury Castle, Croscombe*
2. DHC 1262M/o/E/15, 1774, *Correspondence re Estates (Miles to Fortescue)*, Fortescue Estate
3. Parliamentary Papers, 1802-3, evidence of John Phillis, *Minutes of Evidence on the Woolen-Trade Bill*, Internet Archive
4. SHC DD/C/9, 1901-1934, *Charity Commission records (including extracts from earlier documents)*
5. TNA AR/1044, 1785-1798, records of *Ancient Free and Accepted Masons of England, Lodge of Unanimity*, Wells
6. TNA C202/192/5 180, 1804, *Chancery, Petty Bag Office, Return of Writs, Oaths of Justices of the Peace*
7. SHC A/DUW/1, 1791-5, *Kings Sedgemoor, List of Claims (letter from Billingsley, 1 Jan 1796)*
8. ibid, n 4

Chapter 3 The Wool Trade

1. Tucker, Josiah, 1757, *Instructions for Travellers*, Gloucester
2. Mann, J de L, 1971, *The Cloth Industry in the West of England*, Oxford University Press
3. Parliamentary Papers, 1802-3, evidence of John Phillis, *Minutes of the Select Committee on the Bill respecting the Laws relating to the Woolen-Trade*, Internet Archive
4. Youatt, William, 1837, *Sheep, their Breeds, Management and Diseases*, Baldwin and Cradock
5. Billingsley, John, 1795, *General View..* 2nd ed, p 145
6. TNA IR/26, 1716-1811, *Apprenticeship Records*
7. Morris, M and Morris, M, 1995, *A Complete Transcription of George Horner's Lease Book*, Doulting
8. DHC 1262M/o/E/15, 1774, *Correspondence re Estates (Miles to Fortescue)*, Fortescue Estate

9. *Bath Chronicle,* 20 June, 1776, letter from 'A Friend to the Woollen-Manufacturers'
10. SHC SANHS, 1920, M14-16 [GRN0224424] *Notes and Queries for Somerset and Dorset,* vol 16, re Letters to the War Office, 1776
11. *Norfolk Chronicle,* 12 July 1776, letter from Shepton Mallet dated 2 July 1776
12. Oliver, Andrew, ed, 1972, *The Journal of Samuel Curwen, Loyalist,* Harvard University Press
13. *Bath Chronicle,* 1 Aug 1776, letter in support of Billingsley
14. *Hampshire Chronicle,* 16 Sept 1776, report of a meeting at Bristol of clothiers of the west of England, on 3 Sept 1776
15. ibid n 10
16. *Bath Chronicle,* 10 Oct, 1776, report of a meeting at Bath of clothiers of the west of England, on Oct, 1776
17. ibid, n2, pp 126-7
18. ibid, n 5, p161
19. ibid n 3, pp126-7

Chapter 4 Turnpikes
1. Billingsley, John, 1795, *General View..* 2nd ed, p307
2. Albert, William, 1972, *The Turnpike Road System in England 1663-1850,* Cambridge University Press
3. Acts of Parliament: *Turnpike Roads Act,* 1773, 13 Geo III, c 84
4. Acts of Parliament: *Somerset Roads Act,* 1753, 26 Geo II, c 92
5. Atthill, Robin, 1971, *Old Mendip,* 2nd ed, David and Charles
6. ibid, n 4
7. Acts of Parliament: *Highways Act,* 1765 [Highways and Turnpikes], 5 Geo III, c 38
8. SHC D/T/sm 1776-1873 *Shepton Mallet Turnpike Trust Records*
9. DHC1262M/o/E/15/7, 1779, *Correspondence re Estates, Letter from Billingsley to Fortescue,* Fortescue Estate
10. Acts of Parliament: *Somerset Roads Act,* 1780, 20 Geo III c 85
11. Lambert, Percy, 2009 *Nettlebridge,* Oakhill and Ashwick Local History Group
12. SHC D/T/sm/8 1818-31 *Acquisition of property for road improvement, Ashwick*
13. ibid, n 1
14. Bogart, Dan, 2005, *The Transport Roads of England and Wales,* The Cambridge Group for the History of Population and Social Structure
15. ibid, n 1

Chapter 5 The Oakhill Brewery
1. *Western Gazette,* 16 Aug 1867
2. *The Bath Chronicle,* 14 Mar 1776
3. *The Bath Chronicle,* 24 Feb 1774
4. Combrune, Michael, 1762, *The Theory and Practice of Brewing,* London, J Haberkorn
5. Mathias, Peter, 1959, The Brewing Industry in England, 1700-1830, Cambridge University Press

6. Pryor, Alan, 2015, The Industrialization of the London Brewing Trade, *Brewery History 161*
7. SHC DD/BT/1/59, 1822, *Map of the Estate belonging to Messrs Jillard and Spencer, Shepton Mallet*
8. Phelps, William, 1839 *History and Antiquities of Somersetshire*, p211
9. ibid, n 7
10. Pudney, John, 1971, *A Draught of Contentment, the Story of the Courage Group*, New English Library
11. SHC D/P/she/13/1/11, 1771-77, *Shepton Mallet Poor Rates*
12. Deeds of the Oakhill Inn, held privately
13. ibid, n 2
14. TNA, PRIS 4/5, Mar 1772-Nov 1776
15. *London Gazette*, Issue 11675, 11th June 1776
16. ibid, n 10
17. SHC DD/OB/1, 1776-1818, *Ubley etc, Deeds*
18. SHC Q/REL/40/13, 1767-1831, *Tithing of Shepton Mallet Tax Records*
19. Latimer, John, 1893, *The Annals of Bristol in the 18th Century*, Bristol
20. Photographer and date unknown, Photograph of Oakhill Cottage, Oakhill and Ashwick Local History Group Archive
21. *Taunton Courier and Western Advertiser*, 22 Nov 1810
22. ibid, n 5
23. *Shepton Mallet Journal*, 9 Aug 1918
24. *Bristol Mirror*, 1 July 1809
25. *Bath Chronicle*, 25 Oct 1810

Chapter 6 Agriculture and the Bath and West – Part 1
1. Billingsley, John, 1797, 'The Improvement of Land Lying Waste', *Transactions of the Royal Society of Arts*, vol 15 p184, eBooks: books.google.co.uk
2. BRO 0038/1/1, 1777-91, *Minutes of Annual and Ordinary Meetings of the Bath and West Society* (in this chapter all further references to Minutes are from this record)
3. Hudson, Kenneth, 1976, *The Bath and West, a bi-centenary history*, Moonraker Press
4. Billingsley, John, 1780, 'Account of the Culture of Carrots and thoughts on Burnbaiting on the Mendip-hills', *Letters and Papers of the Bath and West Society*, vol I, Crutwell
5. Billingsley, John, 1783, 'On the Profit of Carrots and Cabbages', *Letters and Papers of the Bath and West Society* vol II, Crutwell
6. Billingsley, John, 1784, 'Culture, Expenses, and Produce of Six Acres of Potatoes', *Letters and Papers of Bath and West Society* vol III, Crutwell
7. *Reading Mercury* 18 April 1785
8. Rack, Edmund, 1784, Editorial, *Letters and Papers of Bath and West Society*, vol III, Crutwell
9. SHC A/AQP/39, 10 Jan 1787, *Letter from Rack to Collinson*

Chapter 7 Agriculture and the Bath and West – Part 2

1. BRO 0038/1/2, 1791-1802, Minutes of the General and Annual Meetings, *Bath and West Society* (in this chapter all further references to Minutes are from this record until noted otherwise)
2. Atthill, Robin, 1957, *An Agricultural Pioneer: the story of John Billingsley*, in the *Western Gazette*, 15 Feb 1957
3. Billingsley, John, 1795, 'A Particular Return of an Experiment made in Sheep-Feeding', *Letters and Papers*, vol VII, article XVI pp 352-60, Bath and West Society, Crutwell
4. Chevalier de Monroy, 1796, 'Foreign Agriculture, or an Essay on the comparable advantage of oxen for tillage in comparison with horses', letter to Sir John Sinclair, *Annals of Agriculture*, vol 31
5. *Annals of Agriculture* 1796, vol 27
6. *Monthly Review*, Sept 1796, vol 21
7. Billingsley, John, 1797, 'The Improvement of Land Lying Waste', vol 15, p184, *Transactions of the Royal Society of Arts*, eBooks: books.google.co.uk
8. BRO 0038/1/3, 1802-11, Minutes of the General and Annual Meetings, *Bath and West Society*
9. Moger, Mrs W, 31 Dec 1955, Letter to Robin Atthill, (including photographs)
10. BRO 0038/1/12/, 1803-17, Minutes of the Superintending Committee, *Bath and West Society*
11. BRO 0038/13/8/1, Two photographs of the Bedfordian medal, *Bath and West Society*
12. Billingsley, John, 1805, 'Remarks on the Utility of the Bath and West of England Society', *Letters and Papers*, vol X, Bath and West Society
13. Billingsley, John, 1807, 'An Essay on the Cultivation of Waste Lands', *Letters and Papers*, vol XI, Bath and West Society

Chapter 8 Enclosures

1. Billingsley, John, 2nd ed, 1795, *General View..*, pp 297-8
2. Williams, Michael, 1976, 'Mendip Farming: the last three centuries', in Atthill, Robin, ed, 1976, *Mendip: a New Study*, David and Charles
3. Young, Arthur, 1808, *General Report on Enclosures*, Board of Agriculture
4. ibid n 1, p 49
5. DHC 1262M/o/E/15/3, 1760, *Map of Ashwick Down*
6. ibid n 3, p 61
7. SHC Q/RDE, 1720-1913, *Somerset Enclosure Acts*
8. Orwin, C, Bonham-Carter, V ed, and Sellick, Roger J, 1997, *The Reclamation of Exmoor Forest*, Exmoor Books
9. Buchanan, B.J, 1982, 'The Financing of Parliamentary Waste Land Enclosure: Some Evidence from North Somerset, 1770-1830', *Agricultural History Review*, vol 30
10. Billingsley, John, 1797, 'The Improvement of Land Lying Waste', *Transactions of the Royal Society of Arts*, vol 15 p172
11. *Bath Chronicle*, 1794, Auction of Land at Little Green Ore, Lot III
12. Billingsley, John, 1805, 'Remarks on the Utility of the Bath and West of

England Society', *Letters and Papers,* vol X, Bath and West Society
13. SHC Q/RDE/27, 1785, *Shepton Mallet Enclosure Act*
14. Williams, Michael, 1976, 'Mendip Farming: the Last Three Centuries', in Atthill, Robin, ed, *Mendip: a New Study,* David and Charles
15. Phelps, Rev William, 1839, *The History and Antiquities of Somersetshire,* vol II, p7
16. SHC Q/RDE/58, 1776, *Doulting and Stoke Lane Enclosure Act*
17. Jefferson, Joyce, ed, 2006, *A History of Stoke St Michael,* Stoke History Group
18. SHC T/Ph/dcl/8, 1790, *Survey of Manor of Shepton Mallet*
19. Billingsley, John, 1798, 'On the Uselessness of Commons to the Poor', *Annals of Agriculture* vol XXXI
20. Eden, Sir Frederick Morton, 1797, *The State of the Poor* vols 1-3
21. ibid, n 3
22. Bingham, V.E, 'Papers of Miss Bingham, Oakhill (1916-1997)', *Private Collection*

Chapter 9 Land Acquisition
1. DHC 1262M/o/E/15/6, 1770-1783, *Accounts and Correspondence, Somerset Estates*
2. ibid, n 1
3. ibid, n 1
4. SHC DD/WG/50/1, 1773-1800, *Vol of Copies of Enclosure Awards, including Ubley (1773)*
5. SHC DD/OB/1, 1776-1818, *Ubley etc, Deeds*
6. SHC DD/OB/2, 1760 on, *Conveyances re Hazel Manor*
7. SHC DD/OB/3, 1818-44, *Ubley, Chewton Mendip Deeds, Manor of Hazel etc*
8. ibid, n 6
9. Atthill, Robin, 1971, *Old Mendip* 2nd ed, pp 13-17, David and Charles
10. ibid, n 7
11. ibid, n 7
12. *Bath Chronicle,* 7 Nov, 1811
13. ibid, n 7
14. Jefferson, Joyce, ed, 2006, *A History of Stoke St Michael,* p12, Stoke History Group
15. SHC T/PH/dcl/8, 1790, *Survey of Manor of Shepton Mallet*
16. DHC 1262M/o/E/15/14, 1791, *Book of Reference to Map of Manor of Ashwick*
17. DHC 1262M/o/E/15/15c, 1803, *Particulars of Manor of Ashwick for Sale*
18. Billingsley, John, 1797, 'The Improvement of Land Lying Waste', *Transactions of the Royal Society of Arts,* vol 15
19. Bath and West Society, 1805, *Editorial, Letters and Papers,* vol. X

Chapter 10 Navigable Canals
1. Clew, Kenneth R., 1985, *The Kennet and Avon Canal,* David and Charles
2. Atthill, Robin, 1971, *Old Mendip* 2nd ed, David and Charles, p48
3. Clew, Kenneth R., 1970, *The Somersetshire Coal Canal and Railways,* David and Charles

REFERENCES

4. Parliamentary Act, *Newbury to Bath Canal Act*, 1794 (34 Geo III) c 90
5. *Bath Chronicle*, 4 Oct 1792
6. *Bath Chronicle*, 7 Feb 1793
7. Billingsley, John, 1794, *General View of the Agriculture in the County of Somerset*, pp 30-2
8. Parliamentary Act, *Somerset to Bradford Canal Act*, 1794 (34 Geo III) c 86
9. ibid n 3
10. Allsop, Niall, 1988, *The Somersetshire Coal Canal Rediscovered*, Millstream Books
11. ibid n 3 pp 164-5
12. Parliamentary Act, *Somerset Canal Act*, 1796 (36 Geo III) c 48
13. Parliamentary Act, *Somersetshire Coal Canal Act*, 1802 (42 Geo III) c 35
14. Clew, Kenneth R., 1971, *The Dorset and Somerset Canal – an Illustrated History (Inland Waterways History Series)*, David and Charles
15. ibid n 14,
16. Parliamentary Act, *Dorset and Inland Navigation Act*, 1796 36 Geo III), c 47
17. Hadfield, E.C.R. 1942, *The Economic History Review*, vol 12, no 1 /2, pp59-67
18. ibid, n 14, p 21
19. *Bath Chronicle*, 16 Oct 1800
20. ibid n 2, p167
21. *Salisbury and Winchester Journal*, 14 Mar 1803
22. *Bath Chronicle*, 10 Jan 1793
23. *Bath Chronicle*, 30 Oct 1794
24. *Bath Chronicle*, 5 Mar 1796
25. *Bristol Mirror*, 8 Sept 1810

Chapter 11 Mining
1. Billingsley, John, 2nd ed, 1795, *General View of the Agriculture in the County of Somerset*, p16
2. ibid n 1, p 27
3. Down, CG and Warrington, AJ, 1971, *The History of the Somerset Coalfield*, Radstock Museum
4. SHC DD/RM/22
5. TNA C/13/632/33, 1807, Court of Chancery, *Waldegrave v Billingsley*
6. Gould, Shane, 1999, *The Somerset Coalfield*, SIAS Survey no 11, Somerset Industrial Archaeological Society
7. ibid n 4
8. ibid n 1, p 20
9. SHC Q/RDE/23, 1800, *Chewton Mendip Enclosure Awards*
10. ibid n 1, p 21
11. ibid, n 1, pp20-1
12. ibid n 1, pp 23-6
13. ibid n 1, p 23
14. SHC DD/X/ATR/2, 1801-5, *Accounts kept by JS Tudor with J Billingsley for Earl Waldegrave*
15. TNA, 1808, Close Roll Index C33/557

Chapter 12 Water Management and Draining the Levels

1. BRO 0038/1/4, 1811-1823, *Committee Minutes, Annual and Ordinary Meetings*
2. Historic England, 2018, *Water Meadows: Introductions to Heritage Assets*, Historic England
3. SHC GRN0311466, 1783, *Letters and Papers,* Bath and West Society, vol. II, p 87
4. Atthill, Robin, undated (probably 1950's), *Sketch of Billingsley Water Meadow,* Downside Archaeological Society
5. Billingsley, John, 1798, *General View of the Agriculture in the County of Somerset,* 2nd ed, p264 [further references in this chapter to Billingsley's work all come from this source unless otherwise noted]
6. Williams, Michael, 1970, *The Draining of the Somerset Levels*, Cambridge University Press.
7. https://wikishire.co.uk/wiki/File:Map_of_Somerset_Levels.png (Creative Commons Licence)
8. Young, Arthur, 1771, *A Farmers Tour of the South and West of England*
9. SHC D/RA/13/1/1 1698-1877, *Somerset Drainage Acts*
10. SHC DD/WY/18/3/8, 1800-01, *Copies of River Brue Drainage Act*
11. SHC D/RA/1/10/4, 1802, *Bound Copy of the Drainage Act, for Axe Drainage*
12. Parliamentary Papers, 1775, *Journals of the House of Commons,* vol xxxiii, 649
13. SHC DD/WY/18/3/5, 1791, *Printed Copy of King's Sedgemoor Enclosure Act*; and D/RA/1/8/1, 1795, *King's Sedgemoor Enclosure Award*
14. SHC A/DUW/1, 1791-95, *King's Sedgemoor List of Claims, as corrected by Commissioners* [John Billingsley's working copy]
15. SHC DD/H1/317, 1794-5, *Henry Hippisley-Coxe, Letters*
16. SHC D/PC/cur.n/5/1/1, 1800, *North Curry, Stoke St Gregory and West Hatch Inclosure Award*

Chapter 13 Ashwick Grove

1. Collinson, Rev John, 1791, *The Antiquities and History of Somerset*
2. Ashwick Grove Estate, 1937, *Auction Catalogue,* Local History Group collection
3. SHC DD/FS/92/12, 1791, *Map of Manor of Ashwick*
4. DHC 1262M/0/E/15/6, *Survey of Manors of Ashwick and Kilmersdon*
5. DHC 1262M/0/LSO/Emborough/2 [actually relating to Ashwick] 1702-31, *Fosse House Tenement*
6. DHC 1262M/0/E/15/7, 1763-1808, *Particulars of Manor of Ashwick*
7. DHC 1262M/0/E/15/14, 1791, *Book of Reference to Map of 1791, Manor of Ashwick*
8. Atthill, Robin, 1971, *Old Mendip* 2nd edn, David and Charles, pp 45/6
9. SHC D/RA/1/8/3, 1776-1877, *Kings Sedgemoor Enclosure* [documents possibly compiled by Robert Granville]
10. SHC DD/SH/71, Strachey, Richard Charles, 1892, *History of Ashwick Grove*
11. ibid n 8, p 45
12. British Library ref ADD 3367, 1820's, *Rev John Skinner Print of Ashwick Grove*
13. DHC DD/BR/py/4, 1803, *Ashwick Deed, Lease for Lands called Ashweek Grove*
14. *Bath Chronicle,* 8 Jan 1814

15. ibid, n 2

Chapter 14 A General View of Agriculture..
1. Billingsley, John, 1794, *General View..*.
2. Sinclair, Rev John, 1837, *Memoirs of the life and work of Sir John Sinclair, Bart*
3. ibid, n 2
4. Holmes, Heather, 2012 and 2013, *Sir John Sinclair: The County Agricultural and the Collection and Dissemination of Knowledge, 1793-1817 Parts I and II,* Journal of the Edinburgh Bibliographical Society
5. Sinclair, Sir John, 1796, *Account of the Origin of the Board of Agriculture and its progress after three years,* Bulmer. [via Holmes]
6. SR RASE B/X/III [MERL], 1795-6, *Board of Agriculture, Letter Book*, p126
7. ibid, n 6
8. Billingsley, John, 1795, *General View*, 2nd edition
9. ibid n 8
10. ibid n 8
11. Marshall, William, 1808, *Review and Abstract of the County Reports of the Board of Agriculture,* vol 2
12. Acland, Sir Thomas Dyke, 1850, *The Farming of Somersetshire,* Journal of the Royal Agricultural Society of England, vol XXIII
13. Wilmot, Sarah, 1996, *The Scientific Gaze: Agricultural Improvers and the Topography of South West England*: in, Brayshay, Mark, ed., *Topographical Writers in South-West England,* University of Exeter Press
14. Billingsley, John, 1795, ibid. n 8, p314

Chapter 15 An Essay on Waste Lands
1. Billingsley, John, 1807, *An Essay on Waste Lands, in Letters and Papers,* vol XI, pp 3-93, Bath and West Society [all following quotations in this chapter are from this source unless otherwise noted]
2. Williams, Michael, 1976, 'Mendip Farming: the last three centuries', in *Mendip: a New Study,* ed Atthill, Robin, David & Charles
3. BRO 0038/1/3, 1802-1811, *Minutes of the General and Annual Meetings,* Bath and West Society
4. Bryant, P, Bennett, S, Collins, T, 2002, *The Story of the Bath and West,* Bath and West of England Society
5. ibid, n 2
6. Osborne, Pip, personal communication
7. Acland, Sir Thomas Dyke, 1850, *The Farming of Somersetshire*, vol XXIII, p726, Journal of the Royal Agricultural Society of England
8. ibid, n 2
9. ibid n 1, p 48
10. Marshall, William, 1796, *The Rural Economy of the West of England,* p 51
11. Atthill, Robin, ed. 1976, *Mendip: a New Study,* David and Charles
12. Neeson, JM, 1993, *Commoners: commoners rights, Enclosures and social change,* Cambridge University Press

Chapter 16 Earl Waldegrave's Steward
1. TNA C13/641/4 ,1807, Countess Waldegrave v Billingsley
2. TNA C13/632/33, 1807, Earl Waldegrave v Billingsley (Revived and Supplementary 1811)
3. TNA C33/550 and C33/555, 1807, Close Roll Indexes
4. *St James' Chronicle, London*, 12 Dec 1807; and *English Chronicle and Whitehall Evening Post*, 10 Dec 1807
5. TNA, C33/557, 1808, Close Roll Index
6. The *Sun, London*, 11 Apr 1808
7. TNA C13/641/4, 1808, Earl Waldegrave v Billingsley, George Pope, John Saunders Tudor and John Taylor
8. TNA C33/579, 1808, Close Roll Index
9. SHC DD/X/ATR/2, 1801-5, *Accounts kept by JS Tudor with J Billingsley for Earl Waldegrave*
10. TNA C13/649/36, 1809, Earl Waldegrave v Billingsley, (Revived and Supplementary 1811)
11. TNA C13/522/13, 1811, Earl Waldegrave v Billingsley, (Revived)
12. TNA C33/587, 1811, Close Rolls Index

Chapter 17 Last Years and Legacy
1. BRO 0038/1/4, 1811-23, Bath and West Society, *Annual and Ordinary Meetings*
2. Historic England, Listed Entry: 1395191
3. BRO 0038/6/1/2, 1796-1848, *Records of the Bath and West Society, Letter Book*
4. SRC DD/OB/2, 1760-1818, *Chewton Mendip etc Deeds* [containing an office copy of Billingsley's Will of 23 Dec 1807]
5. *Taunton Courier*, 25 Oct,1810
6. *Bath Chronicle*, 3 Oct, 1811
7. Tablet in St James' Church, Ashwick
8. TNA PROB11/1527, 1807, *Will of John Billingsley Esq*
9. ibid n 1
10. Hobhouse, Benjamin, 17 Dec, 1811, *Eulogium to John Billingsley at the Annual Meeting of the Bath and West of England Society*; reported in: MEASE, James, 1813, *Archives of Useful Knowledge*, Philadelphia, vol III, p 175
11. Hudson, Kenneth, 1976, *The Bath and West, a bi-centenary history*, Moonraker Press
12. BRO 0038/6/2, no date, *Letter Book, Records of the Bath and West of England Society*
13. *Gentleman's Magazine*, Oct 1811, volume 81, part 2, p395
14. *Bath Journal*, 30 Sept 1811
15. SHC DD/OB/3, 1818-44, *Ubley, Chewton Mendip Deeds, Manor of Hazel etc*
16. *Bath Chronicle*, 17 Oct, 1811
17. *Bath Chronicle*, 31 Oct, 1811
18. *Bath Chronicle*, 8 Apr, 1813
19. Phelps, W, 1839, *The History and Antiquities of Somerset*, vol 2, p127
20. Acland, Sir Thomas Dyke, 1850, *The Farming of Somersetshire*, Journal of the Royal Agricultural Society of England, vol XXIII, p726

21. Atthill, Robin, 1964, *Old Mendip*, David & Charles
22. Atthill, Robin, 1957, *An Agricultural Pioneer: the story of John Billingsley*, in the *Western Gazette*, 15 Feb 1957
23. BRO 0038, 1805, 'Remarks on the Utility of the Bath and West of England Society', *Letters and Papers,* vol X, Bath and West Society p37

Selected Bibliography

Albert, W, 1972, *The Turnpike Road System in England, 1663-1840*, Cambridge University Press
Atthill, Robin, 1964, *Old Mendip*, David & Charles
Atthill, Robin, ed, 1976, *Mendip, A New Study*, David & Charles
Boughey, Joseph and Hadfield, Charles, 2007, *British Canals: the Standard History*, Sutton Publishing
Brayshay, Mark, ed, 1996, *Topographical Writers in South-West England*, University of Exeter Press
Bryant, Philip, Bennett, Susan, Collins, Ted, 2002, *The Story of the 'Bath and West', innovation and application*, The Royal Bath and West of England Society
Chambers, J D and Mingay, G E, 1966, *The Agricultural Revolution, 1750-1880*, B T Batsford
Clew, Kenneth R, 1970, *The Somersetshire Coal Canal and Railways*, Bran's Head Books
Clew, Kenneth R, 1971, *The Dorset and Somerset Canal*, David and Charles
Collinson, Rev John, 1791, *The Antiquities and History of Somerset*
Down, C.G, and Warrington, A.J, 1972, *The History of the Somerset Coalfield*, David and Charles
Gould, Shane, 1999, *The Somerset Coalfield*, SIAS Survey no 11, Somerset Industrial Archaeological Society
Hudson, Kenneth, 1976, *The Bath and West, a bi-centenary history*, Moonraker Press
Mann, J de L, 1971, *The Cloth Industry in the West of England, from 1640-1880*, Oxford University Press
Mathias, Peter, 1959, The Brewing Industry in England, 1700-1830, Cambridge University Press
Mingay, Gordon E, 1997, *Parliamentary Enclosure in England: an Introduction to its Causes, Incidence and Impact – 1750- 1850,* Routledge
Murch, Jerome, 1890-1, *The History and Literature of the Bath and West of England Society*, Journal of the Bath and West and Southern Counties Society, vol 10
Overton, Mark, 1996, *The Transformation of the Agrarian Economy 1500-1850*, Cambridge University Press
Ponting, Kenneth G, 1971, *The Woollen Industry of South-West England*, Adams and Dart
Porter, Roy, 1982, *English Society in the Eighteenth Century*, Pelican Books
Thorley, Lin and others, 2017, *The Changing Face of Ashwick Parish*, Oakhill & Ashwick Local History Group
Ward J R, 1974, *The Finance of Canal Building in 18th Century England*, Oxford University Press

Williams, Michael, 1970, *The Draining of the Somerset Levels,* Cambridge University Press

Acknowledgements

Writing this book has often seemed like ploughing a lonely furrow, but now at the end I realise there are so many people I need to thank. Not everyone who helped is mentioned by name, nonetheless they are appreciated.

First, thank you to those at Hobnob, Dr John Chandler and Dr Louise Ryland-Epton, for your constructive advice and patience; and to Robyn Hawkins for making the index. The many staff at archive centres have all been extremely helpful and professional, thank you to those at: the South West Heritage Trust at Taunton and at Exeter; Bath Guildhall Record Office; Bristol Archives; the Museum of English Rural Life; Downside Abbey Archives; and The National Archives at Kew. Thank you to the following for their kind permission to publish images: the Victoria Art Gallery (Bath and North East Somerset Council) for the image of the John Billingsley portrait by Joseph Hutchinson; to the Somerset Heritage Centre at Taunton for images of their records; to Canon Thomas Atthill for images of his father Robin's work; and to Percy Lambert for an image of his sketch.

Several professionals and individuals with specialist knowledge have read sections and commented and/or added useful information, at times correcting my mistakes; including Prof Richard Hoyle, David Brown, Derrick Hunt, Shane Gould, Michael McGarvie, Mick Davis and Ben Mackay. All have busy lives so I am very grateful to you for taking the time and trouble to engage. Several Mendip farmers have commented on the text and/or in discussions; among them Bruce Luffman, Chris Norman, Sam Matthews and Amanda Alvis. Many thanks for your interest in the project and taking time out to share your expertise.

My thanks, too, for the friendship and encouragement of those involved in local history in my area: the Oakhill and Ashwick Local History Group and the Shepton Mallet Local History Research Group. In particular Matt Brookbank, Alan Stone, Percy Lambert, Thomas Atthill, and Amanda Miles – thank you all for answering numerous questions, producing information and/or reading sections.

From all the above I would like to single out for special mention Bruce Luffman and Matt Brookbank. Both have gone above and beyond, spending more time helping me than I had any right to even hope for. Thank you so much.

Finally, but most important, my thanks to my husband David, for your love, support and encouragement.

Index

Abbey Church House 59
Acland, Sir Thomas Dyke, 157, 165, 192
Agriculturalist(s) 11, 54, 144, 147, 149, 159, 161, 192-5
Ailsebury (Aylesbury), Earl of 85
American Revolution, 3
Ammerdown, 185
Ancient Laws of Mining 179
Anglican Church 5
Annals of Agriculture 63
Anstice, Robert 126
Argyle Buildings 136, 183
Ashwick 1-8, 13, 17, 23-5, 27, 30, 33, 34, 37, 46, 86,
Ashwick Down 4, 76
Ashwick Grove 1, 4, 6, 7, 9, 11-14, 25, 26, 28, 30, 31, 42, 73, 83, 88, 92-94, 102, 116, 127, 130-41, 142, 148, 182-5, 188-91
Ashwick Manor 93
Ashwick Presbyterian Chapel 11
Ashwick St James 5-8, 10, 13, 14, 185
Atlantic 2, 77
Atthill, Robin 25, 59, 62, 91, 105, 136, 137, 190, 192, 193
Avon, River 97, 98
Axe, River 118, 120-4, 129, 149

Banker(s) 14, 44, 45, 101
Banks, Sir Joseph 62, 64
Bartley, Nehemiah 65, 67-9, 184
Basset, Sir Francis 128
Bath, Marquess (Marquis) of 64, 177
Bath (City of) 1, 2, 12-14, 21, 23-26, 29, 40, 49-50, 54, 58, 59, 65-7, 73, 98-103, 112, 118, 136, 138, 149, 183-5, 188, 190
Bath and West Society 11, 13, 22, 44, 48-73, 75, 81, 115, 120, 158-62, 171, 184-9
Bath Chronicle 20, 39, 40, 46, 49, 58, 99, 104, 139, 185, 191
Bath Guildhall 190
Bath Journal 188
Bath, Lord 55
Battle of Waterloo 174
Beauchamp 109

Beckington 20
Bedford, Duke of 66
Bedfordian (Bedfordean) Gold Medal 14, 66, 68, 72, 158, 184, 190
Beeches, The 38, 42
Belgium 37
Bell Inn 43
Berkshire Canal 101
Billingsley family (members other than John Billingsley junior) 13, 33, 130, 135, 183
Billingsley, John senior 7, 9, 10, 13, 17, 88, 133, 134
Billingsley, Marianne 13, 14, 184, 186, 189, 190
Billingsley, Mary (nee Greening) 7, 9, 13, 133
Billingsley, Mary (nee James) 5, 7, 13,
Billingsley, Mary (nee Wells) 11, 13, 186-8
Billingsley, Mary Wells 11, 14
Billingsley, Nicholas 7-8, 9, 10, 13, 88-9, 133-4
Billingsley, Rev Nicholas junior 5-7, 10, 13, 17, 136-8
Billingsley, Rev Nicholas senior 5-7, 131-2, 137
Billingsley, Richard 5
Billingsley, Rev Samuel 6, 9
Billingsley's publications – *see* Publications
Bishop of Bath and Wells 112
Black Swan 45
Blake, George 39-41, 45
Blandford 103
Board of Agriculture 12, 14, 55, 63, 65, 125, 142-57, 192
Bolingbroke, Lord 124
Bonvill, Lord of Charterhouse 112
Bowl Down 56
Bowles, John 13, 17, 92
Bradford Canal Act 100
Breeding 60, 63, 69, 153
Brendon Hills 117
Brent Marsh 149
Breweries (*see also* Oakhill Brewery) 33, 35, 37, 39, 42, 44
Brewers 14, 34, 37, 39, 40, 42, 44, 45, 46

Brewing 33-47
Bridgewater Canal 96
Bridgwater 31, 106, 126
Bridgwater Bay 118, 149
Brighton 184
Bristol 5, 21, 24, 37, 41, 42, 45, 49, 65, 80, 97, 98, 102, 105, 106, 112, 149
Bristol and Taunton Canal 97, 105, 106
Broderip, Edmund (also, Broderib and Brodrib) 92, 185
Brown, Mr 56
Brue, River 120, 121-124, 129, 149
Brue Drainage Act 124
Burge, Messrs 133
Burnham 124
Burton Pynsent 128
Butleigh 121, 128

Caisson-Lock 101
Camden Crescent 184, 185
Camden, Earl of 184
Camden Place 136, 184, 185
Camerton 108
Cary, River 118, 124, 126
Castle Hill 4, 89
Chard 22, 149
Chard Canal 106
Charity Commission 10
Charterhouse 111, 112
Cheddar 77, 79
Chevalier de Monroy 63
Chewton 112, 113
Chewton Mendip 12, 14, 78, 80, 91, 95, 111, 164, 165, 173, 175
Church Rates 134
Clandown 110
Clerk 43, 127, 128
Cloth (see Wool)
Clothier(s) 2, 7, 10, 13, 15-22, 25, 26, 33, 48, 88
Coke, Thomas 48
Coleford 17
Coles, Mr 176
Collieries 31, 99, 104, 109, 110
Colliery 104, 109, 110
Collinson, Rev John 6, 42, 49, 58, 124, 130, 135, 137
Commissioner(s), of Enclosure 11, 14, 74-87, 93, 109-21, 125-8
Commissioner of Turnpikes 25, 26, 28, 30
Commoner(s) 75, 79, 85, 86, 120, 125, 128, 161, 194
Compton Martin 90, 91, 94
Congresbury Yeo 118
Cook, John 22

Cookow Thorn 91
Cotswold sheep 60, 61
Cottager(s) 83-5, 161
Court cases 12-14, 172-81, 185, 188, 195
Court of Chancery 174, 179, 186
Court of Equity 178
Court de Wick 62
Court of Sewers 119, 125
Cowl Street Chapel 10
Croker, Mr 146
Crook Mr 60
Cropping 169-171
Croscombe 79
Curwen, Samuel 20

Davis, Thomas 57, 67, 152, 153, 180
Day and Masters map 1, 91
Devon 4, 6, 11, 13, 71, 160
Devonshire 151
Dickens, Charles 100, 181
Dinder 79
Dirrick, Phillip 90
Dishley 61
Dorchester 19, 37
Dorset(shire) 16, 37, 49, 55, 61, 78
Dorset and Somerset Canal 97, 102, 108
Down and Warrington 110
Downside 5
Drainage – see Somerset Levels
Dressing 15
Duchy of Cornwall 35, 82, 84, 109
Duke of Bedford 66
Dunbald-Clize (Clyse) 126
Dundas Aqueduct 101
Dundas, Hon Charles 98
Dundon 122
Dunkerton 100
Dye 17

Earl Waldegrave – see Waldegrave
Earl Fortescue – see Fortescue
East Allington 11
East Anglia 14, 50
East Harptree 79, 94, 111-13, 166, 173, 175-9,
East Harptree Manor 112, 173, 175-9
East Horrington 121
Eastman 40
Eden, Sir Frederick 86
Elliott, Mr 128
Elverstoke Manor 93
Employment 3, 15, 18, 86, 107, 114, 138, 161, 162-4, 179, 186
Enclosures 3, 4, 11, 12, 14, 35, 65, 72, 74-87, 90, 91, 109, 111, 119-22, 125, 127-

INDEX 211

9, 143, 144, 149, 151, 158-62, 166-7, 170-1, 175, 180, 182, 192-4
Enclosures on Mendip 74, 89, 93
England 3, 15, 19, 21, 23, 46, 48, 75, 96, 115, 118, 145
Essay on Waste Lands – *see* Publications
Essex 49, 57
Eton 174
Europe 3, 75
Eveleigh, John 184
Exeter 10
Exeter Coach 135
Exmoor 16, 78, 116, 118

Fairchild, William 124
Fanshaw, Miss 66
Farmer Sully 59
Farmers 48, 52, 54-6, 59, 61-2, 64, 65, 69, 79, 83, 84, 128, 144, 147,148, 152-57, 159, 163,165-71, 195
Farmer's Magazine 52
Farming 2, 10, 11, 28, 49-51, 56, 68, 72, 75, 76, 92, 148, 149, 158, 162-6, 169, 175, 177, 178, 201, 202
Fernhill Farm 91
Ford, E.H. 138
Forest of Mendip 2, 35, 53, 75
Forsyth, William 138
Fortescue, Dorothy 11
Fortescue family 4, 8, 11, 14, 89, 93, 132
Fortescue, Earl 14, 26, 79, 82, 89, 93, 132-5
Fosse Estate 139, 191
Fosse House Tenement 6, 88, 89, 93, 132, 133, 135
Foster, Dr James 6, 137, 138
France 46
Frankpledge 90
Freemasons 11, 14
French Revolution 3, 104
Frome (Froome) 9, 19, 21, 23, 42, 44, 104, (149)
Fromefield 42
Fussell 104

Gardener(s) 138, 139
General Enclosure Act 63, 76
General View of Agriculture (*see also* Publications) 142-57
Gentleman's Magazine, 188
George, Phillip 41
George's Brewery 37, 41, 42
Georgian 136, 184
Gloucester(shire) 15, 34, 49, 172
Goul, Nathaniel 'Jab' 86
Gould 110

Government 20, 24, 34, 45, 96, 104, 142
Grame, Mr Thomas 180, 186
Grand Western Canal 98, 105
Green Ore 12, 14, 65, 80-2, 94, 95, 115, 163
Greening, Mary 7, 9, 13
Green's Wood 132, 138
Greylake Fosse 124
Groovers/grooving) 112

Hales, Sir Philip 125
Hamworthy 103
Hanley, Sylvanus 28
Harden, Joseph 20
Harridge Wood 4, 138
Harrowing 53, 54, 80, 165, 170
Hazel Manor (and Estate) 11, 14, 90-2, 94-5 111, 173, 185, 191
Helps, James 20
Herschel, William 64
Hetling House 59, 65
High Littleton 108
Highbridge 122
Highland Clearances 142
Highways Act 23
Highway Tax 31, 154
Hippisley-Coxe, Henry Esq 99
Hippisley, Sir John Cox 186, 187
Hippisley-Coxe, Sir John 127
Hobhouse, Sir Benjamin 44, 64, 66, 187, 188
Holcombe 86
Holmes 143, 146
Horner, Thomas 14, 92
Horrington 79, 121
Hosier (hozier) 2, 7, 15
House, Joseph 27, 28
Hudson, Kenneth 51, 187
Hues, Robert 191
Huntspill 121
Husbandry 55, 153, 162, 187, 192

Ilchester, Earl of 49, 50, 125
Insolvent Debtor 40
Investor(s) 23, 24, 41, 45, 96, 98, 101, 103, 106, 126, 159
Ireland 4, 14, 63
Irishman 138
Isle of Wight 190

James, Charles 7
James family 132
 James, Mrs Mary 39
 James, Samuel (of Ocle) 89
 James, William 7
Jenkins, Richard 110

Jenkins, William 17
Jenkyn's 191
Jenner, Edward 64
Jessop, William 100, 126
Jillard family 35-46
　Jillard, Rev Peard 44
　Jillard, Robert Haskoll 45
　Jillard, William Peard 44, 46
　Jillard, William Vernon 45
　Jillard and Spencer 35, 37, 38
Jolliffe, Thomas 92, 185
Jordan, James 13, 14, 33-35, 39-44, 82, 90, 92
Jordan, Mrs 186
Joy, Lord 134
Judge(s) 16, 54, 61
Judgement 59, 156, 176-8, 180
Justice of the Peace 12, 14, 190

Kaposvar Room 190
Keepers Cottage 138
Kennet and Avon Canal 97, 98, 100, 103
Kennet, River 98
Kew 138
Keynsham 59
Kilmersdon 5, 92
King, Annie 174
King George III ('Farmer George') 62-4
King, Mr 190
Kingsbridge 13
King's Bench Prison 40
King's Sedgemoor 14, 120, 121, 124-6, 129, 149
Knitters 2, 17

Labour 16, 18, 22-4, 30, 31, 85, 87, 124, 161, 168
Labourers 2, 113, 108, 162
Lady Chatham 128
Lady Day 191
Lambert, Edward 28
Lambing 168
Land Agent, 4, 18
Landlord 152, 155
Landowner(s) 3, 4, 49, 74-9, 97, 99, 102, 103, 107, 110, 120, 122, 124-8, 143, 144, 160, 194
Langman, Joseph 28
Lapland 2
Laws and Orders of the Mendip Miners 150
Leasehold 82, 89, 93, 95
Leasing 4, 8, 81, 93, 109
Legal Cases – *see* Court cases
Leicester (sheep) 60-2, 190
Levels, the – *see* Somerset Levels

Liming 81, 165
Litigation 14, 180, 188
Locke, Richard 3, 124, 125
Lodge of Unanimity 11, 14
Logmill 18, 92
Logwood Mill (also Bloodwood) 13, 17
London 14, 31, 40, 45, 98
Long Cross 25
Longhouse Ground 191
Longleat 55, 59
Lord Bath 55
Lord Bolingbroke 124
Lord Joy 134
Lord(s) of the Soil 112
Lord Rivers, Lord of the Manor of Shillingstone 103, 104
Lords Royal (of Mineries) 112-13
Lord Somerville 72
Lord's Share 113
Lovell, Mayor John 80
Ludlow's Mine 109

Mackrell 40
Maitland, John 45
Maltsters 37, 45
Marchants Hill 28
Marksbury 190
Marshall, William 145, 156, 157, 168
Masbury 10, 95
Master 15, 17, 57
Matthews, William 58, 59, 65, 68, 69, 71, 95
Meare 122
Mells 92, 103, 104
Mendip (and Mendip Hills) 1-4, 11, 16, 23, 24, 30, 31, 35, 37, 41, 53, 62, 65, 70, 75, 77-81, 84, 86, 87, 89, 116-18, 121, 128, 135, 150, 151, 160, 163-5, 167, 169-71, 192, 195
Merchants 15, 46
Merino (sheep) 16, 62-4, 198
Middle Pit 109, 110
Midford 59, 100
Midlands 86, 96
Midsomer Norton 109
Miles, William 4, 134, 135
Milling 15
Mills 15, 16
Milton, Mr 66
Mindery Batch 113
Miners 80, 108, 113, 150
Mining 2, 4, 11, 75, 77, 94, 96, 107-14, 115, 149, 169, 170, 173, 178, 179, 181, 182
Minister(s) 4, 5, 6, 9, 44, 169
Mitchell, Thomas (senior and junior) 90, 91
Moger, Mrs W (nee Mary Kathleen Fortescue

Weigall) 54, 67, 190
Mogg, Jacob 100
Moggridge 34
Monkton Combe 100
Monmouthshire 99
Moorwood Pits 2, 109
Murch, Sir Jerom 5

Navigable canals (*see also* individual canals) 96-106
Nettlebridge 26-9, 102-5, 108, 116
New Zealand 190
Newbury 98
Nonconformist/(ism) 4-12, 16, 33, 45, 46, 95
Norfolk 49, 56, 57, 59
Norfolk Chronicle, 19
Norris 199
North Curry 127, 128
Norton Down 56
Norwich 53

Oakhill 1, 3, 30, 35-47, 97, 194
Oakhill Brewery 1, 3, 4, 10, 13, 14, 22, 33-47, 49, 82, 89, 92, 93, 95, 102, 103, 108, 115, 117, 185,186, 194
Oakhill Cottage 42-3
Old Down 25, 99
Ottley, Thomas 68
Oxford 23

Padget, William 27, 111
Park Farm 137
Parliament 23, 24, 26, 27, 63, 74-6, 96, 98, 114, 120, 123, 142, 159
Parrett, River 118, 124, 126
Parry, Dr Caleb Hillier 64
Parsons family
  Parsons, Isaac 9, 13, 135
  Parsons, Mary (nee Billingsley) 9, 13, 90
  Parsons, Thomas 9, 14, 90, 91, 135, 185
  Parsons, William 9, 185
Parsons, Mr White 69, 70
Paulton 100
Pensford 108
Perkins family
  Perkins, Joanna 33
  Perkins, Dr Joseph 33
  Perkins, Richard 33, 34, 40, 41, 105, 185
Phelps, William 36, 192
Pickwick, Eleazer 100-2
Pickwick, Moses 100
Pilton 121
Pitcot 109
Pitt, William (Prime Minister) 128, 142

Ploughing(s) 14, 53-7, 59, 60, 65, 66, 80, 129, 152, 154, 164, 167, 170, 192, 193
Pondsmead 43, 138
Poole 103
Poor Rate(s) 3, 18, 84, 86, 134, 151
Pope, Alexander 59, 138
Pope, George 174
Pratt, Charles 184
Preacher 4-6
Presbyterian(s) 5, 6, 10, 11, 43, 95, 134
President 45, 51, 67, 69, 195
Priddy 77, 79, 80
Priddy Minery 112
Priestly FRS, Dr Joseph 49
Prince of Wales 101
Prince Regent 184
Prince William, Duke of Gloucester 172
Prizewinning 12, 80, 157, 192
Publican 34, 45
Publications by Billingsley
  'On the Profit of Carrots and Cabbages, Letters and Papers of the Bath and West Society' 53
  'Results on Fifteen Acres of Scotch Cabbage' 55
  'Account of the Culture of Carrots and thoughts on Burn-baiting on the Mendip-Hills' 53
  'Culture, Expenses, and Produce of Six Acres of Potatoes, Bath and West Society' 54
  'Remarks on the Utility of the Bath and West of England Society, with an Account of the Progress of Improvement in the County of Somerset' 70-1
  'A Particular Return of an Experiment made in Sheep-Feeding' 60-2
  'On the Uselessness of Commons to the Poor' 14, 84-6
  'A Recapitulation of the Hints for Improvement' 150-7
  General View of Agriculture of Somerset, with Observations on the Means of its Improvement, 1st and 2nd editions, 12, 14, 22, 31, 42, 50, 55, 63, 72, 73, 77, 78, 82, 101, 107, 111, 117, 123, 125, 127, 142-57, 158, 171, 182, 193
  'An Essay on the Cultivation of Waste Lands' 14, 72, 73, 82, 158-71, 193
  'Essay on the Improvement of Land Lying Waste, RSA (1797) 12, 14, 64, 80, 81
Publow 9, 90
Pye Hill 30

Quaker(s) 45, 49, 59, 161

Quantock Hills 117
Quarrying 2, 107

Rack, Edmund 14, 49, 50, 57, 58, 60
Radstock 99, 100, 102, 109-11, 173, 175
Radstock Manor 109
Radstock Old Pit 109
Rennie, John 98, 100
Reynolds, Mr Nicholas 43
Ricards (Rickards), Robert 69
Rivers, Lord, Lord of the Manor of Shillingstone 103, 104
Robins, Thomas 56
Rodney Stoke 77, 79, 121
Roman Fosseway 1, 25, 131
Royal Agricultural Society 192
Royal Board of Agriculture 157,
Royal Dragoons (Troop of) 19-21
Royal Marriages Act 172
Royal Society for the Encouragement of Arts (also, Royal Society and Royal Society of Arts - RSA) 12, 14, 48, 64, 80, 94

St Cuthbert Out (Wells) 79, 80, 95, 121
Salisbury, Earl of 49
Salters Hall Synod 6
Saunders, John (Tudor) 113, 179
Scotland 54, 145
Scottish Highlands 142
Secretary 49, 50, 51, 54, 59, 60, 65, 69, 133, 143, 156, 161
Sedgemoor 121
Servant(s) 52, 138, 139, 140, 172, 176,
Seymour, Arthur Penrose 189, 190
Seymour, Henry Fortescue 190
Seymour, Rev George Penrose 189, 190
Shannon, 2nd Earl of 14, 63
Share(s)/Shareholder(s) 96, 100, 109-10
Shearing 15, 64
Sheep 2, 14, 16, 26, 55, 60-3, 64, 66, 69, 118, 129, 142, 152, 169, 170, 171
Sheppey, River 16
Shepton Mallet 1, 3, 5, 7, 10, 13, 15, 16, 19-21, 25, 26, 33, 35, 39, 40, 42, 44, 45, 52, 84, 89, 93, 95, 131, 133, 134, 135, 191
Shepton Mallet Caravan 138
Shepton Mallet Enclosure Act 82-4
Shepton Mallet Journal 45
Shepton Mallet Manor 35, 87
Shepton Mallet Trust 11, 24
Shillingstone 103, 104
Shillingstone Manor 103
Sinclair, Sir John 142, 143, 145, 147
Skinner, Rev John 136-137
Smallcombe Coal Company 109-10

Smallcombe Colliery/Pit 109
Smeaton 100
Smith, William 'Strata' 34, 100
Solicitor 92, 185
Somerset Coal Canal 97, 99-103, 110
Somerset Coalfield 2, 3, 37, 99, 102, 110, 108, 110
Somerset Levels (Drainage of) 2, 12, 14, 77, 79, 94, 115-29, 134, 163, 188, 193
Somersetshire 100
Somerville, Lord 72
South Marsh 125, 149
Southampton 138
Southdown 61, 190
Southwark 40
Spencer Family 35, 45, 46
 Spencer, Frederick 45
 Spencer, John 45, 46
 Spencer, John Plummer 45
 Spencer, John Maitland 45
Spinning (spinners) 13, 15, 18, 20, 21, 22
Spinning Jenny 18, 20, 22
Stallard, William 28
Stephens, James 99
Steward (*see also* Waldegrave and Bath) 12,
Stocking Makers 2, 7, 16, 17
Stogdon, Hubert 6
Stoke Bottom 17, 116
Stoke Lane Manor 11, 14, 18, 83, 92, 95, 115, 116, 191
Stoke St Michael (previously Stoke Lane) 7, 11, 14, 17, 35, 83, 92, 95, 99, 115-16, 131, 133
Ston Easton 4, 99
Stour, River 104
Strachey family 136, 137, 140, 191
Stratton 28
Strode, John Esq 20
Stroudwater Canal 97
Suffolk 57
Supervisory (also Superintendence) Committee (of the Bath and West Society) 14, 66, 68
Surgeon 34
Surveyor (s) 23, 31, 76, 106, 122, 124. 126, 145
Sutcliffe, John 100
Swann Inn 83, 105

Tamworth 56
Tapp, John 30
Taunton 106
Taunton Assizes 178
Taylor, John 179
Tetbury 56

Thames 98
Thurlow, Lord Chancellor 177, 178
Timsbury 100
Tollgatherers 24, 27, 28
Tolls 23-28, 99, 101
Tone, River 118
Travellers 23, 25, 30
Trowbridge 16
Tucker, Josiah 15
Tudor, JS (John Tudor Saunders) 113, 179
Tudway, Clement 80
Tudway, Robert 109
Tugwell, Mr 56
Turner, Canon John 127-8
Turnpikes 11, 13, 23-32, 80, 154, 194
Turnpike Roads Act 24
Turnpike Trusts 11, 13, 23
Twickenham 138
Twigge, Miss 186

Ubley 11, 13, 41, 53, 89-92, 94, 111, 185
Upper Camden Place 184

Vagg, Mr Henry 56
Vancouver 151
Vicar 11
Vice President (of Bath and West Society) 51, 52, 66, 73, 187, 189, 190
Victoria Art Gallery 190

Waldegrave family 12, 14, 95, 109-13, 172-81, 185
  Waldegrave, Elizabeth Laura, Countess (also Lady Waldegrave and Dowager Countess Waldegrave) 110, 172-6, 178, 186
  Waldegrave, George, 4th Earl, 172
  Waldegrave, George, 5th Earl, 172
  Waldegrave, John James, 6th Earl 12, 14, 78, 95, 99, 109-13, 127, 164
Wales 34, 99
Walker, James 111, 175
Walpole 174
War Office 18, 19, 21
Wareham 103
Warminster 19, 21
Warrington 110
Water Meadows 12, 14, 93, 115-17, 138, 191

Watt, James 110
Weaver(s) 15, 16, 18-20, 40
Wedmore 85
Weigall, Mary Kathleen Fortescue (*see* Mrs W Moger)
Weldon, Robert 101,
Wells (City of) 1, 24, 25, 79, 80, 105, 106, 112, 121, 122, 127
Wells Cathedral 127
West Camel 69
West Harptree 77, 79, 94
West of England 15, 21, 22, 55
West Riding, Yorkshire 15
West Sedgemoor 127
Western Canal (*see also* Grand Western Canal) 98
Weymouth, Lord 59
Whalley, Francis 109
Whalley, Rev Richard Chaple 109
Whitbreads 44
White, Edward 45, 46
White Hart Inn 58
White Horse and the Moon (Public Houses) 42
White, Miss Lydia 14, 138-9, 183
White Post 26
White, William 122, 123, 126, 150
Whitworth, Robert 100
Wigmore Farm 164-65, 173
Wilkins Esq, Charles 92
Williams, Michael 117, 163
Wiltshire 15, 49, 55, 59, 61, 71, 78, 99, 101
Wincanton 103
Winchester 154
Wool Trade (including wool/woollen trade and cloth/industry) 2, 3, 5, 7, 10, 15-22, 33, 40, 44, 51, 65, 88, 89, 92, 194
Workhouse 18
Wraxall 190
Wright, Mr James Camper 180

York House 49
Yorkshire 15
Youatt 16
Young, Arthur 23, 54, 75, 76, 86, 119, 124, 143, 145, 156, 157, 171, 194
Young, William 106

www.ingramcontent.com/pod-product-compliance
Lightning Source LLC
Chambersburg PA
CBHW040256170426
43192CB00020B/2822